# Key Concepts of Polymerase Chain Reaction

# Key Concepts of Polymerase Chain Reaction

Edited by **Giorgio Salati**

New York

Published by Callisto Reference,
106 Park Avenue, Suite 200,
New York, NY 10016, USA
www.callistoreference.com

**Key Concepts of Polymerase Chain Reaction**
Edited by Giorgio Salati

International Standard Book Number: 978-1-63239-440-8 (Hardback)

# Contents

Preface     VII

Chapter 1   **Study of Mycobacterium Tuberculosis**
**by Molecular Methods in Northeast Mexico**    **1**
H. W. Araujo-Torres, J. A. Narváez-Zapata,
M. G. Castillo-Álvarez, MS. Puga-Hernández,
J. Flores-Gracia and M. A. Reyes-López

Chapter 2   **Measuring of DNA Damage by Quantitative PCR**    **14**
Ayse Gul Mutlu

Chapter 3   **Detection of *Apple Chlorotic Leaf Spot Virus***
**in Tissues of Pear Using *In Situ* RT-PCR**
**and Primed *In Situ* Labeling**    **26**
Na Liu, Jianxin Niu and Ying Zhao

Chapter 4   **Analysis of Genomic Instability and Tumor-Specific Genetic**
**Alterations by Arbitrarily Primed PCR**    **40**
Nikola Tanic, Jasna Bankovic and Nasta Tanic

Chapter 5   **Analysis of Alternatively Spliced Domains**
**in Multimodular Gene Products - The Extracellular Matrix**
**Glycoprotein Tenascin C**    **58**
Ursula Theocharidis and Andreas Faissner

Chapter 6   **Lack of Evidence for Contribution of eNOS, ACE and AT1R**
**Gene Polymorphisms with Development of Ischemic Stroke**
**in Turkish Subjects in Trakya Region**    **72**
Tammam Sipahi

Chapter 7   **Overview of Real-Time PCR Principles**    **86**
Morteza Seifi, Asghar Ghasemi, Siamak Heidarzadeh,
Mahmood Khosravi, Atefeh Namipashaki, Vahid Mehri Soofiany,
Ali Alizadeh Khosroshahi and Nasim Danaei

Chapter 8    **PCR Advances Towards the Identification of Individual and**
             **Mixed Populations of Biotechnology Microbes**                    **124**
             P. S. Shwed

Chapter 9    **Submicroscopic Human Parasitic Infections**                     **136**
             Fousseyni S. Touré Ndouo

Chapter 10   **Detection of Bacterial Pathogens**
             **in River Water Using Multiplex-PCR**                            **152**
             C. N. Wose Kinge, M. Mbewe and N. P. Sithebe

Chapter 11   **Identification of Genetic Markers Using Polymerase Chain**
             **Reaction (PCR) in Graves' Hyperthyroidism**                     **176**
             P. Veeramuthumari and W. Isabel

Chapter 12   **PCR-RFLP and Real-Time PCR Techniques**
             **in Molecular Cancer Investigations**                           **190**
             Uzay Gormus, Nur Selvi and Ilhan Yaylim-Eraltan

             **Permissions**

             **List of Contributors**

# Preface

This book aims to highlight the current researches and provides a platform to further the scope of innovations in this area. This book is a product of the combined efforts of many researchers and scientists, after going through thorough studies and analysis from different parts of the world. The objective of this book is to provide the readers with the latest information of the field.

The primary objective of this book is to illustrate recent concepts in molecular biology with focus on the application to human, animal and plant pathology, in numerous aspects like analysis, prevention and treatment of diseases, prognosis, etiology and use of these methodologies in comprehending the pathophysiology of several diseases that impact living beings. The topics included are measuring DNA damage; overview of PCR principles; submicroscopic human parasitic infection and real time PCR techniques in cancer investigations. This book will be beneficial for researchers, students and professionals engaged in the field.

I would like to express my sincere thanks to the authors for their dedicated efforts in the completion of this book. I acknowledge the efforts of the publisher for providing constant support. Lastly, I would like to thank my family for their support in all academic endeavors.

**Editor**

# 1

# Study of Mycobacterium Tuberculosis by Molecular Methods in Northeast Mexico

H. W. Araujo-Torres[1,3], J. A. Narváez-Zapata[2], M. G. Castillo-Álvarez[1],
MS. Puga-Hernández[4], J. Flores-Gracia[5] and M. A. Reyes-López[1,*]

[1]*Conservation Medicine Lab., Centro de Biotecnología Genómica del
Instituto Politécnico Nacional, Cd. Reynosa, Tamps,*
[2]*Industrial Biotechnology Lab., Centro de Biotecnología Genómica
del Instituto Politécnico Nacional, Cd. Reynosa, Tamps,*
[3]*Centro de Investigación en Ciencia Aplicada y Tecnología Avanzada
del Instituto Politécnico Nacional, Altamira, Tamps,*
[4]*Laboratorio Estatal de Salud Pública de Tamaulipas, Cd. Victoria, Tamps,*
[5]*Instituto Tecnológico de Ciudad Victoria, Cd. Victoria, Tamps*
*México*

## 1. Introduction

One third of the world population is afected by TB and one million people did die this year 2011 in undeveloped countries (Venkatesh et al., 2011). In Tamaulipas, a Northern State of Mexico and a border state between USA and Mexico, frequency is 26.9 new TB cases per 100,000 people, twice of national rate of 12.85 cases per 100,000 people (Ferrer et al., 2010). Only on the border of Tamaulipas about 320 cases are diagnosed each year. Many of these cases correspond to people from other states of Mexico, probably by geographic position and by migration problematic of this study zone (Fitchett et al., 2011). Only 92% of the treated population are cured mainly because much of these people are poor and whose nutritional status directly affects the possibility of quick recovery (SSA, 2009).

The long presence of this disease has increased the need to know specifically which *Mycobacterium tuberculosis* strains are circulating in the region. Additionally, it is necessary to know the antibiotic/susceptibility profile of these strains since many of them acquire resistance against the traditional antibiotics along time.

In general, the diagnostic of this disease is traditionally conducted by using gold standard techniques focused to identify the presence of *M. tuberculosis* in clinical specimen of humans or cattle. These techniques included the strain of microorganism in Ziehl-Neelsen and culture in Lowenstein-Jensen medium (Cadmus et al., 2011), both regarded as reference techniques in the diagnosis of TB. Differentiation among mycobacteria of the *M. tuberculosis* complex (MTC) and other than MTC (NMTC) is accomplished by applying biochemical tests: niacin production, catalase activity, thermostable at 68 ° C and reduction of nitrate.

---

* Corresponding Author

Actually, detection for *M. tuberculosis* has been shorted due mainly to application of molecular methods directly to clinical samples. Usually, the detection of this bacterium takes 2 to 4 weeks (Marhöfer et al., 2011). Some molecular techniques are already on the market, being the most commonly used *AMPLICOR M. tuberculosis PCR test* (Roche), *M. tuberculosis* Direct test (MTDT) (GenProbe) and LCX *M. tuberculosis* assay (Abbott). In addition, PCR amplification of ribosomal sequences (Ribotyping) or amplification of repetitive intragenic consensus sequences (i.e. ERIC-PCR, spoligotyping, MIRU-VNTR), among others, are the usual methods used to specifically discriminate among different *M. tuberculosis* strains (Rodwell et al., 2010; Pang et al., 2011).

The use of molecular approach is also applied in the analysis of the antibiotic resistance of these isolates. In this sense, molecular detection of specific mutations in genes involved with drug resistance has successfully been applied in the identification of these (i.e. detection of mutations on *rpo*B, *kat*G, *mab*A, etc.) (Sala & Hartkoorn, 2011).

As an example of the conjunction of the background described above, this chapter briefly presents a work of potential tuberculosis patient samples, to which mycobacteria were isolated to determine whether any of them were resistant to some antibiotics and if it could be grouped by health districts of Tamaulipas and also grouped the isolates strains in MTC and NMTC, and identify them, potentially.

Therefore, the main aim of this study was determinated by using a molecular approach the specific *M. tuberculosis* strains presents in the region and identify specific mutations in these strains related with drug resistance. The information here generated helps to take epidemiological decisions aimed to control and to prevent this disease in the Northeast of Tamaulipas.

## 1.1 TB statistical

This main issue is due to produce a resistance against antibiotics used traditionally to control the disease. The first antibiotic against TB was created in the 40's decade. Consequently, the incidence of this disease declined in the following decades, especially in developed countries. However, in the last 20 years it has been observed an increase in the TB cases around of the world, particularly due to generation of new *M. tuberculosis* strains with resistance of the traditional antibiotics, or multidrug resistance (Yew et al., 2010). The death associated to TB may increase in undeveloped countries since some others diseases as the HIV may duplicate the death frequency in patients with both diseases (Havlir & Barnes 1999; Sonnenberg et al., 2005). Given this situation the TB was declared as global emergency by WHO (WHO 1993).

## 1.2 Situation of TB in northern Mexico

The Northeast state of Tamaulipas exhibits a high peak of occurrence of TB with regarding at Mexican rate and the frequency of this disease has remained stable during the last 10 years (Ferrer et al., 2010). Besides, this region is a natural corridor for exporting of cattle between Mexico and EE.UU. Therefore, both countries have commitment to keep safe their borders (Fitchett et al., 2011). Among the factors that may partially explain it, the high migrations rate reported in this region with people from other Mexican states (mainly

Veracruz, Coahuila, Mexico city, among others) and other Central America countries (i.e. Guatemala, Honduras, among others).

One of the reasons of this migration may be high number of manufacturing factories that offer a high number of jobs and that appeal and wait to travel. In addition, many of these people only remain of 4 to 5 years here and wait to travel to U.S.A. As it was previously mentioned, many of these people are poor and with low nutritional status and their lifestyle (i.e. drug or alcohol consumes) could prompted the TB disease (Wagner et al., 2011). Therefore, TB will become a big issue between Mexico and USA. From here, both countries have agreements on health and security cooperation. These agreements include the fast detection of this disease and the discrimination among *M. tuberculosis* strains (Fitchett et al, 2011).

### 1.3 Multidrug resistant in Mycobacterium tuberculosis

Recently, it has been reported an increase in the TB cases around of the world, particularly due to generation of new *M. tuberculosis* strains with resistance to traditional antibiotics, or multidrug resistance (Sougakoff, 2011). In 2007, the 14th edition of the Merck list shows 30 different anti-TB drugs, many analogues or prodrugs of antibiotics, as the first line of defense against this disease. In Mexico, the antibiotics most commonly used are the rifampicin (RIF), the isoniazid (INH), the pyrazinamide (PZA), the streptomycin (STR) and the ethambutol (EMB) (Borrell, Gagneux., 2011).

Worldwide, rifampicin is the drug mostly in the control of this bacterium (Connell et al., 2011). These antibiotics are not enough to halt the emergence and spread of multidrug resistant (MDR) strains causing a serious problem for the TB control and increasing public health problems (Zumi, et al. 2001). This have prompts the development of fast and reliable diagnostic process to detect, to discriminate, and to evaluate resistance of *M. tuberculosis* strains against main drugs.

### 1.4 Molecular approaches

Molecular biology has allowed detection of DNA or RNA sequence of different mycobacteria. An example of these approaches is using probes. Probes were prepared from nucleic acid sequences complementary to the DNA or RNA sequences from different species (including *M. tuberculosis, M. avium, M. kansasii, M. gordonae.*), which may be labeled with radioactive isotopes (hot probes) or chromogenic substances (cold probes). The gene probe is capable of binding or hybridizing with a homologous fragment of the study sample, which has been previously denatured by physical means. Hybridization of the probe to its complementary fragment is easily detected with addition of a marker. The main advantages of these techniques are fast and specific. Its disadvantages high cost and that many probes cannot identify species within the MTC.

Typing techniques based on amplification of nucleic acids by PCR provide a fast and reliable approach to obtain genetic information about bacteria or microorganism groups. Molecular typing methods for tuberculosis are based on that those infected by strains of *M. tuberculosis* have the same genotype (genetic fingerprinting) and are epidemiologically related, while those infected with different genotypes (unique patterns) are not.

Among the techniques of molecular biology that are currently used, as: ribotyping, the PCR amplification of repetitive extragenic palindromic sequences (REP-PCR) and the repetitive intragenic consensus sequences of *Enterobacteriaceae* (ERIC-PCR). These techniques can also be used in clinical studies to establish patterns of colonization and to identify sources of transmission of infectious microorganisms, which may contribute to a better understanding of the epidemiology and pathogenesis thereby helping to develop disease prevention strategies (Struelens, MESGEM, 1996).

## 1.5 Ribotyping

Ribotyping technique applied in the diagnostic of diseases has been used for differentiation of bacterial serotypes involved with the occurrence of outbreaks. Additionally, this technique has an extended use in the study of nosocomial fungus (Pavlic and Griffiths, 2009). Ribotyping is also used to study the ecology, the genotypic variation and the transmission of Streptococcus mutants from person to person (Alam et al., 1999). The patterns are simplified to ribotyping, making visible the DNA fragments containing parts or all of ribosomal genes, sometimes detected bacterial serotypes (Pavlic and Griffiths, 2009).

Because of the epidemiological and clinical importance of some bacterial strains such as *M. tuberculosis*, it is interesting the application of related techniques like typing by PCR, in breaking through in a better understanding of the ecology and epidemiology of these bacteria. Some studies show this approach to evaluate the discriminatory power of different methods for genotyping of MTC isolates, they compared the performance of i) IS6110 DNA fingerprint, ii) spoligotyping and iii) 24-loci MIRU-VNTR (mycobacterial interspersed repetitive units - variable number of tandem repeats) typing in a long term study on the epidemiology of tuberculosis (TB) in Schleswig-Holstein, the most-northern federal state of Germany (Roetzer et al, 2011), other group studied the clustered cases identified using a population-based universal molecular epidemiology strategy over a 5-year period. Clonal variants of the reference strain defining the cluster were found in 9 (12%) of the 74 clusters identified after the genotyping of 612 M. tuberculosis isolates by IS6110 restriction fragment length polymorphism analysis and mycobacterial interspersed repetitive units-variable-number tandem repeat typing. Clusters with microevolution events were epidemiologically supported and involved 4 to 9 cases diagnosed over a 1- to 5-year period (Pérez-Lago et al, 2011), another study was to compare polymerase chain reaction (PCR)-based methods-- spoligotyping and mycobacterial interspersed repetitive units (MIRU) typing--with the gold-standard IS6110 restriction fragment length polymorphism (RFLP) analysis in 101 isolates of Mycobacterium tuberculosis to determine the genetic diversity of M. tuberculosis clinical isolates from Delhi, North India (Varma-Basil et al 2011) and finally, a study where Forty three isoniazid (INH)-resistant *M. tuberculosis* isolates were characterized on the basis of the most common INH associated mutations, katG315 and mabA -15C→T, and phenotypic properties (i.e. MIC of INH, resistance associated pattern, and catalase activity). Typing for resistance mutations was performed by Multiplex Allele-Specific PCR and sequencing reaction (Soudani et al, 2011).

## 1.6 ERIC-PCR

Amplification of enterobacterial repetitive intergenic consensus by PCR (ERIC-PCR) has only been used sporadically to detect mycobacteria. ERIC sequences are repetitive elements of 126 bp that appear to be restricted only to transcribed regions of chromosome. Its position in the genome appears to be different in different species. As any technique, ERIC-PCR is used as typing, to study the clonal relationship in various Gram-negative bacteria such as *Acinetobacter baumannii*. The DNA patterns obtained with the ERIC-PCR are usually less complex than those generated by other techniques such as REP-PCR. The technique is quick and easy to perform, and provides highly reproducible results (Gillings & Holley 1997.).

Additionally, the presence of ERIC sequences has been detected in genome of *M. tuberculosis* (Sechi et al, 1998). Studies showed that the level of differentiation obtained by ERIC-PCR is superior to that obtained by the RFLP-IS6110 genetic profile comparable to that obtained by (*GTG*) *5-PCR* fingerprinting (PCR-GTG). The use of the PCR-GTG, a repetitive marker in the *M. tuberculosis* chromosome with an IS6110 sequence has been successfully applied to a PCR-based fingerprinting method. This method is fast and sensitive and can be applied to the study of the epidemiology of infections caused by *M. tuberculosis* and therapeutic implications for health, particularly when the IS6110 RFLP-DNA profile does not provide any help.

## 2. Objective

The aim of this study was to conduct a molecular characterization of mycobacteria strains by typing and drug resistance gene mutations from samples of potential TB patients in Northeast Mexico.

## 3. Methods

Two strategies were conducted to study the samples isolated from patients with probable TB clinical diagnosis from the State Public Health Laboratory of Tamaulipas (LESPT) from The State of Tamaulipas, MX. The first one was the identification of strains as belonged or not to MTC. Second one was to detect mutations on the genes related to drug resistance to major antibiotics against *M. tuberculosis.*

### 3.1 Samples and cultures

Specimens included in this study were collected over a period of 16 months (October 2008 to January 2010) from acid-fast bacilli AFB-positive sputum obtained from the State Public Health Laboratory (LESPT). Basically, LESPT concentrates most of the TB cases from Tamaulipas. All the samples were taken under the informed consent of the patients. . In addition, a structured test was used to obtain standard demographic and epidemiologic data of the patients. Two sputum consecutive specimens were collected from each individual. These samples were mixed with 1% cetylpyridinium chloride and immediately transported to the LESPT where they were stored at 4o C (Kent and Kubica, 1985). All strains cultured were identified to species level by standard microbiological procedures in the LESPT.

### 3.2 DNA extraction

Samples were first lysed (tissue samples were mechanically disrupted) and proteins simultaneously denatured in the appropriate lysis buffer. QIAGEN Proteinase K was then added and after a suitable incubation period, lysates were loaded onto the QIAGEN Genomic-tip. DNA binds to the column while other cell constituents passed through. Following a wash step to remove any remaining contaminants, pure, high-molecular-weight DNA was eluted and precipitated with isopropanol. Hands-on time for the complete procedure was just 45 minutes for samples.

Bacterial strains obtained from patients with TB were preliminary analyzed by an antibiogram test to verify if these strains exhibit some class of antibiotic resistant. Approximately, One hundred consecutive strains were selected to further molecular characterization.     Bacteria selected were growth in solid Lowenstein-Jensen and 7H9 Middlebrook broths supplemented with 10% (vol/vol) of oleic acid-albumin-dextrose-catalase. After that, the samples were incubated for at least 8 weeks.  DNA from bacterial samples was obtained from those grew strains by used the QIAGEN kit (QIAGEN) of according to manufacturing instructions

### 3.3 Molecular detection of *M. tuberculosis*

The following primers were used (Yeboah-Manu et al. 2001): spacer region-specific primers, spacer region 33 specific (5'ACACCGACATGACGGCGG3') and spacer region 34 specific (5'CGACGGTGTGGGCGAGG3');     IS6110     (5'GGACAACGCCGAATTGCG'3     and 5'TAGGCGTCGGTGACAAAGGCCAC'3), and Mycobacterium genus-specific TB11 (sequence     5'ACCAACGATGGTGTGTCCAT3')     and     TB12     (sequence 5'CTTGTCGAACCGCATACCCT3').  Expected PCR products are 550, 439, and 172 and 99 bp, respectively.

PCR mixtures contained 20 µl of 2× PCR mix, 10 µL of primer mix with each primer at 0.66 pmol/µL, 0.2 µL of Taq polymerase enzyme (Roche Diagnostics), and 10 µL of extracted DNA. The PCR conditions were 95°C for 3 min; 30 cycles of 95°C for 20 s, 65°C for 30 s, and 72°C for 30 s; and 72°C for 7 min. After PCR, the products were analyzed by electrophoresis in agarose gel.

### 3.4 Typing methods

For ribotyping, the standardization of PCR was done using three sets of primers to amplify the 16S region (Table 1). Note: The primers 16S and 16S R F were used for amplification of 16S Mycobacterium.

| Primer | Sequence | Size (nt) | Tm (°C) | Reference |
|---|---|---|---|---|
| R1 | 5'-TTGTACACACCGCCCGTCA-3' | 19 | 62.3 | Sechi A, et |
| R2 | 5'-GAAACATCTAATACCT-3' | 16 | 46.5 | al, 1998. |
| 16S F | 5'AGAGTTTGATCCTGGCTC-3' | 18 | 57.62 | Strom et al, |
| 16S R | 5'-CGGGAACGTATTCACCG-3' | 17 | 59.61 | 2002. |
| P13P F | 5'-GAGGAAGGTGGGGATGACGT-3' | 20 | 64 | Sorrell et al, |
| P11P R | 5'-AGGCCCGGAACGTATTCAC-3' | 19 | 60 | 1996. |

Table 1. Primers used for amplification of 16S Ribosomal DNA of Mycobacterium.

A set of chosen primers, which amplified for desired sequences are shown below (Table 1). (Strom et al, 2002).

Ribotyping by PCR was performed with two primers complementary to conserved regions. The sequences of the primers were described on Table 1. Amplifications were carried out in a final volume of 25 µl. Twenty five cycles of amplification were performed, with each cycle consisting of 2 min of denaturation at 94°C, 45 seconds of annealing at 62°C, and 1 min at 72°C. The last cycle consisted of a 7 min extension at 72°C. The amplification products were visualized after electrophoresis at 90 V for 90 min in a 2% agarose gel, and the gel was stained with SYBER Gold (Invitrogen).

### 3.5 ERIC-PCR

For ERIC-PCR, a pair of primers (Sechi et al, 1998) used and their characteristics are described below (Table 2).

Amplification reactions were performed in a volume of 50 µl with final amounts of 1 U of Taq polymerase, 20 mM Tris-HCl (pH 8.3), 50 mM KCl, 1.5 mM MgCl2, and 200 µM of deoxynucleoside triphosphate (Gibco, BRL, Life Technology, Paisley, United Kingdom). The reaction mixtures were then incubated for 5 min at 95°C, followed by 35 cycles of 94°C for 30 s, Touch-down (47-57°C), and 65°C for 4 min and a final extension at 70°C for 7 min. The amplification products were visualized after electrophoresis at 90 V for 90 min in a 2% agarose gel, and the gel was stained with SYBER Gold (Invitrogen).

| Primer | Nucleotide sequence | Size (nt) | Tm(°C) |
|--------|---------------------|-----------|--------|
| ERIC 1R | 5'-ATGTAAGCT CCT GGGGATTCAC-3' | 22 | 62.7 |
| ERIC 2 | 5'-AAGTAAGTGACT GGGGTGAGCG-3' | 22 | 64.5 |

Table 2. Primers for ERIC-PCR

### 3.6 Gene drug resistant analysis

Eight pairs of PCR primers (PR1 to PR16) were used to simultaneously amplify regions of eight genes associated with resistance to six antituberculosis drugs. In addition, eight pairs (PR17 to PR32) of internal PCR primers were then used to determine the DNA sequences of these genes (Table 3 and 4)

### 3.7 Sequencing

PCR products obtained from only 36 out of 100 bacterial strains for ERIC-PCR and 15 bacteria drug resistant were purified with an EXO-SAP. Components were supplemented with gold buffer (Applied Biosystem) and sequenced on an Applied Biosystem 310 Genetic analyzer (ABI Prism 310 Genetic analyzer), using big dye terminator cycle sequencing Ready Kit (Applied Biosystem).

For Drug resistant, the purified samples were analyzed with the ABI PRISM 3100 Genetic Analyzer (Applied Biosystems). The DNA sequences are collected and edited with Data Collection software version 1.01 and Sequencing Analysis version 3.7 (Applied Biosystems) and compared with those of M. tuberculosis H37Rv (GenBank access no. NC_000962) with the program Geneious version 4.5.4 (Software Development Biomatters Ltd).

Additionally, for ribotyping, sequence of 16S Ribosomal DNA of mycobacterial determined in an ABI Prism® 3130. Obtained sequences were analyzed in NCBI database using BlastN analysis (http://www.ncbi.nlm.nih.gov/blast/Blast.cgi). Alignment editing of 16S sequences was performed by Chromas Lite 2.0, BioEdit Sequence Alignment Editor Version 7.0.4.1 and CLC Sequence Viewer Version 6.1 software.

Finally, the comparison between isolated Mycobacteria and reported Mycobacterium tuberculosis strains were made by a microbial identification and phylogenetic analysis of obtained data using MEGA4 Software Version 4.0.2. Tree topologies were determined by methods of Minimum evolution criterion and Maximum Parsimony, with a value of reliability, "Bootstrap" of 100 replications for phylogenetic analysis.

## 4. Results and conclusion

Male population in Tamaulipas is the most affected by TB with 61% of the isolates evaluated in this work. Geographical distribution of infected people represented a greater proportion in Central and South of the State with 52% and 45% of isolates evaluated, respectively.

| Region | Sequences | Position | Size (bp) |
|--------|-----------|----------|-----------|
| rpoB | PR1 (forward) 5-CCGCGATCAAGGAGTTCTTC-3<br>PR2 (reverse) 5-ACACGATCTCGTCGCTAACC-3 | 1256–1275<br>1570–1551 | 315 |
| katG | PR3 (forward)<br>5-GTGCCCGAGCAACACCCACCCATTACAGAAAC -3<br>PR4 (reverse) 5-TCAGCGCACGTCGAACCTGTCGAG-3 | 1–32<br>223–2200 | 2,223 |
| mabA | PR5 (forward) 5-ACATACCTGCTGCGCAATTC-3<br>PR6 (reverse) 5-GCATACGAATACGCCGAGAT-3 | -217 a -198<br>1145–1126 | 1,362 |
| embB | PR7 (forward) 5-CCGACCACGCTGAAACTGCTGGCGAT-3<br>PR8 (reverse) 5-GCCTGGTGCATACCGAGCAGCATAG-3 | 640–665<br>3387–3303 | 2,748 |
| pncA | PR9 (forward) 5-GGCGTCATGGACCCTATATC-3<br>PR10 (reverse) 5-CAACAGTTCATCCCGGTTC-3 | -80 a -61<br>590–572 | 670 |
| rpsL | PR11 (forward) 5-CCAACCATCCAGCAGCTGGT-3<br>PR12 (reverse) 5-GTCGAGAGCCCGCTTGAGGG-3 | 4–23<br>575–556 | 572 |
| rrs | PR13 (forward) 5-AAACCTCTTTCACCATCGAC-3<br>PR14 (reverse) 5-GTATCCATTGATGCTCGCAA-3 | 428–447<br>1756–1737 | 1,329 |
| gyrA | PR15 (forward) 5-GATGACAGACACGACGTTGC-3<br>PR16 (reverse) 5-GGGCTTCGGTGTACCTCAT-3 | 1–19<br>397–379 | 398 |

Table 3. Primers for multiplex-PCR (Sekiguchi et al. 2007)

Thirty-seven out of 40 samples were analyzed by 16S gene sequences, 34 of them were grouped in the MTC, and the 3 remaining sequences were integrated into the NMTC (Figure 1). Moreover, from 37 sequences analyzed, only 12 of these showed polymorphisms on a segment of 250 nucleotides with an average size of 750 nucleotides for each sequence. Of these 12 sequences, only 3 isolates were grouped in NMTC, the 9 remaining isolated strains show polymorphism in their nucleotide sequences belong to the MTC. Two sequences of isolates tested showed 100% and 98% identity respectively with the species of *M. fortuitum* according to our analysis in the NCBI database (strain *M. fortuitum* 16S gene with accession number DQ973806.1 and strain *M. fortuitum* 16S gene with accession number AY457066.1, respectively).

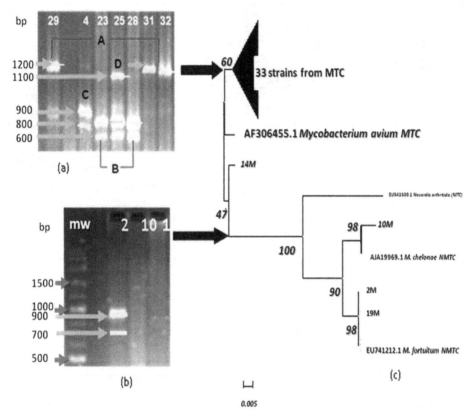

Fig. 1. Comparison of genetic profiles from isolated mycobacteria by ERIC-PCR VS Ribotyping (a) In picture, letters A, B, C, and D show four different genetic profiles of ERIC-PCR grouped in 7 isolates of mycobacteria. Arrow on direction of figure (c) indicates 7 isolated strains are part of the MTC. (b) Image shows ERIC-PCR amplifications of 3 isolates (samples 2M, 10M, 19M) clustered in NMTC. (c) Phylogenetic tree based on comparison of 16S gene mycobacterium species from MTC and NMTC. *Mycobacterium spp* sample.

A third sequence of one isolated strain showed 100% identity with *M. chelonae* (*M. chelonae* strain T9 with AM884324.1 access number). In this sense, it is useful to mention that LESPT identified these strains as *M. tuberculosis* based only on their microbiological and biochemical results. It is important to mention that LESPT does not conducted molecular analysis to identify their samples.

One out of the two identified strains was *M. fortuitum*, made by sequencing, but no for microbiology, since this was identified as *M. tuberculosis* by LESPT. The other isolated itself coincided with both techniques.This gives us a different result, although microbiology diagnostic and taken at this time, we do believe strongly that this is due to *M. fortuitum*.

| Gene | Sequences | Position |
|------|-----------|----------|
| rpoB | PR17 5-TACGGCGTTTCGATGAAC-3      (complementary strand) | 1529–1512 |
| katG | PR18 5-ACGTAGATCAGCCCCATCTG-3 (complementary strand) | 689–670 |
|      | PR19 5-GAGCCCGATGAGGTCTATTG-3 | 574–593 |
|      | PR20 5-CCGATCTATGAGCGGATCAC-3 | 1162–1181 |
|      | PR21 5-GAACAAACCGACGTGGAATC-3 | 1729–1748 |
| mabA | PR22 5-ACATACCTGCTGCGCAATTC-3 | -217 a -198 |
| embB | PR23 5-ACGCTGAAACTGCTGGCGAT-3 | 646–665 |
|      | PR24 5-GTCATCCTGACCGTGGTGTT-3 | 1462–1481 |
|      | PR25 5-GGTGGGCAGGATGAGGTAGT-3 (complementary strand) | 1596–1577 |
|      | PR26 5-CACAATCTTTTTCGCCCTGT-3 | 2007–2026 |
|      | PR27 5-GCGTGGTATCTCCTGCCTAAG-3 | 2581–2601 |
| pncA | PR28 5-GGCGTCATGGACCCTATATC-3 | -80  -61 |
| rpsL | PR29 5-CCAACCATCCAGCAGCTGGT-3 | 4–23 |
| Rrs  | PR30 5-CAGGTAAGGTTCTTCGCGTTG-3 (complementary strand) | 979–959 |
|      | PR31 5-GTTCGGATCGGGGTCTGCAA-3 | 1291–1310 |
| gyrA | PR32 5-GATGACAGACACGACGTTGC-3 | 1–19 |

Table 4. Primers for sequencing (Sekiguchi et al. 2007)

The use of the reference strain H37Rv of *M. tuberculosis* sequence served as a reference or guide for clustering of the isolates studied, since in the phylogenetic analysis, the type strain H37Rv was integrated with the 34 isolated MTC, thus reaffirming the phylogenetic relationship of isolates tested with the species *M. tuberculosis* (Figure 1). On the other hand, the reference sequence of *Nocardia arthritidis* (No. Access EU841600) showed no relation with the 37 evaluated strains in phylogenetic analysis, being totally separated from the two complexes formed, MTC and NMTC. This comparison is done, because of *Nocardia* is also acid-resistant and can be confused with *Mycobacteria* on microscopic analysis, hence the importance of making the comparison. Then, It should be mentioned that the identification of mycobacteria isolated from the 16S sequences was proved to be an appropriate strategy to establish the level of genetic relatedness among isolates studied and know how related isolates were isolated or if these could be separated into MTC and NMTC.

ERIC-PCR technique gave 4 different genetic profiles for mycobacteria (Figure 1). It should be emphasized that three of these genetic profiles are consistent with those reported in molecular epidemiology studies by amplifying sequences ERIC (Sechi et al, 1998). From 34 isolates clustered in the complex of M. tuberculosis, seven of these 4 make up the genetic profiles obtained by ERIC-PCR. Then they expect the rest of the isolates (27 isolates) that make up this complex terms grouped in the 4 genetic profiles (A, B, C and D) obtained by ERIC-PCR.

However, one aspect to consider in the results obtained in this work is that of obtaining the 4 different genetic profiles amplified by ERIC-PCR, they did not allow to discriminate among species of MTC and NMTC as it was expected, those profiles or genetic patterns of 3 isolated strains have been totally different from other profiles of 7 isolated strains (Figure 1), all of 10 isolated strains are in MTC.

Regarding to genetic relatedness of 40 isolates of mycobacteria studied, phylogenetic analysis of 16S gene sequences showed 37 sequences, which formed two groups. The first group of MTC consisting of 34 isolates and the second group resulted in NMTC consisting of 3 isolates. The percentages of identity were from 98% to 100% for isolates clustered in both complexes.

The analysis on relationship between the isolates studied and their geographical origin revealed that the Mycobacterium tuberculosis complex is distributed both in the central and south of the state of Tamaulipas, MX. Meanwhile, species such as *M. fortuitum* and *M. chelonae* are only found circulating in the central region of Tamaulipas.

It should be noted that the isolates studied are only samples originating from the central and southern Tamaulipas. No isolated strains were obtained from northern part of Tamaulipas, which would have complemented the results of this research. For example, as mentioned before in those border states (US-Mexico border) there is a great number of people and cattle to move from different parts of country and abroad, which could suggest existence of different strains of *M. tuberculosis* in Tamaulipas, and even the presence of other species found in the central and southern Tamaulipas, MX, allowing us to know not only that isolated strains in each region are preferably, but also known how those isolated strains are circulating all around Tamaulipas.

Note this work was limited to samples of LESPT; however the proposal will be to analyze samples from all health districts in Tamaulipas, and analyze samples of other states, such as Veracruz and/or Coahuila. Additionally, we will seek to refine the ERIC-PCR and implementing the MIRU-VNTR and spoligotyping for more complete diagnosis. In addition, arrangements are made between the County LESPT and McAllen, Texas, United States, to have samples (about 14,000 isolates) identified and stored in the United States from Tamaulipas.

Finally, the preliminary results were shown (Table 5), where mutations, insertions, transversions, and transitions were found. In general, the mutations obtained did not alter the chemical or structural composition of proteins that confer resistance to an antibiotic to the mycobacteria and their regions sequenced. In these particular cases, we selected to work with isolated strains were resistant to antibiotics commonly administered in Mexico, the results obtained for the case of pyrazinamide, a silent mutation was found, so that the resistance exhibited by the bacteria should be caused by mutations on the sequenced region. In the case of isoniazid *mabA* gene, we found an insertion within the gene that could be the cause of resistance exhibited.

These results indicate that DNA sequencing-based method was effective for detection of MDR strains. However, when novel mutations in drug resistance-related genes are detected by the method, it is essential to also perform drug susceptibility testing, because novel mutation may not be associated with drug resistance.

| Gene | Number of sample | Changes | |
|------|------------------|---------|---|
| | | Nucleotide | Protein |
| *rpo*B | 2 | CGG-TGG | R476W |
| *rrs* | 2 | TGG-AGG | W193R |
| *rps*L | 2 | AAA-AAG | K121K |
| *pnc*A | 2 | GGT-GGC | G75G |
| *mab*A | 2 | 702 T insertion<br>-15 C-T upstream | |
| *gyr*A | 2 | GAG-CAG | E21Q |

Table 5. Relationship of changes found in the sequences of the genes of interest.

## 5. Conclusion

In conclusion, two strategies were carried out to study the samples isolated from patients with TB diagnostic of LESPT from Tamaulipas, MX. The first was the identification of isolates and determine if these isolates belonged or not to MTC. Second, to determinate if mutations in primary sequences of genes related to resistance to major antibiotics used to kill mycobacteria in Tamaulipas, could be detected.

For the first part of the study, there were used 3 strategies, a multiplex-PCR, ERIC-PCR, and ribotyping. For the second direct amplification of 16S DNA region was performed.

Multiplex-PCR for 99% of the samples coincided with the microbiological results, identifying *M. tuberculosis*, primary. In the case of ERIC-PCR, the samples could be grouped into 4 different groups; however it could differentiate between MTC and NMTC. Finally, ribotyping produced promising results by discriminating the isolated strains and identifying 99% as *M. tuberculosis*.

Finally, the results indicate that DNA sequencing-based method was effective for detection of MDR strains. However, when novel mutations in drug resistance-related genes are detected by the method, it is essential to also perform drug susceptibility testing, because novel mutations are not always associated with drug resistance.

## 6. Perspectives

This kind of work will answer other questions: is it necessary ribotyping before or after ERIC-PCR or multiplex-PCR and it is important to recognize each species of Mycobacteria to understand if TB strains would circulate all around Tamaulipas and if those ones would be or get in USA too? In a few years we will understand this phenomenon; meanwhile this chapter makes the first approach to understand how TB strains are moving and if those strains are or not drug resistant on a border State between USA and Mexico. The present investigation continues, pending to sequence regions of resistance to pyrazinamide and ethambutol, which are largest genes.

## 7. Acknowledgments

Financing of this research was by Instituto Politécnico Nacional (projects SIP 20090679 and SIP 20100504). Narváez-Zapata J. A. and Reyes-López M. A. are fellows of CONACYT-SNI,

COFAA, and EDI from Instituto Politécnico Nacional. Authors are also grateful for the support received from Network of Drug Development and Diagnostic Methods (RED FARMED) from CONACyT. Finally, authors really appreciated the samples submitted for QFB. Norma Alicia Villareal Reyes from LESPT.

## 8. References

Alam S., Brailsford S., Whiley R. & Beighton D. (1999). PCR-Based Methods for Genotyping Viridans Group Streptococci. Journal of Clinical Microbiology. Vol. 37, No. 9. p. 2772-2776.

Borrell S, Gagneux S. (2011). Strain diversity, epistasis and the evolution of drug resistance in Mycobacterium tuberculosis. Clin Microbiol Infect. Jun;17(6):815-20

Cadmus SI, Falodun OI, Fagade OE. (2011). Methods of sputum decontamination with emphasis on local tuberculosis laboratories. Afr J Med Med Sci. 40(1):5-14

Connell DW, Berry M, Cooke G, Kon OM. (2011).Update on tuberculosis: TB in the early 21st century. Eur Respir Rev. 20(120):71-84

Ferrer G, Acuna-Villaorduna C, Escobedo M, Vlasich E, Rivera M. (2010). Outcomes of multidrug-resistant tuberculosis among binational cases in El Paso, Texas. Am J Trop Med Hyg. 83(5):1056-8

Fitchett JR, Vallecillo AJ, Espitia C. (2011). Tuberculosis transmission across the United States-Mexico border. Rev Panam Salud Publica. 29(1):57-60

Gillings, M. and Holley, M. (1997). Repetitive element PCR fingerprinting (rep-PCR) using enterobacterial repetitive intergenic consensus (ERIC) primers is not necessarily directed at ERIC elements. The Society for Applied Bacteriology, Letters in Applied Microbiology. No. 25. p. 17-21.

Havlir, D.V. and Barnes, P.F. (1999) Tuberculosis in patients with human immunodeficiency virus infection. N. Engl. J. Med. 340, 367-373

Kent, P. T., and G. P. Kubica. (1985). Public health mycobacteriology: a guide for the level III laboratory. U.S. Department of Health and Human Services, Centers for Disease Control, Atlanta, Ga.

Marhöfer, RJ, Oellien F, Selzer PM. (2011). Drug discovery and the use of computational approaches for infectious diseases. Future Med Chem. 3(8):1011-25

Pang Y, Zhou Y, Wang S, Lu J, Lu B, He G, Wang L, Zhao Y. (2011). A novel method based on high resolution melting (HRM) analysis for MIRU-VNTR genotyping of Mycobacterium tuberculosis. J Microbiol Methods. 86(3):291-7

Pavlic M, Griffiths MW. (2009). Principles, applications, and limitations of automated ribotyping as a rapid method in food safety. Foodborne Pathog Dis. 6(9):1047-55

Pérez-Lago L, Herranz M, Martínez Lirola M; on behalf of the INDAL-TB Group, Bouza E, García de Viedma D. 2011. Characterization of Microevolution Events in Mycobacterium tuberculosis Strains Involved in Recent Transmission Clusters. J Clin Microbiol. 49(11):3771-3776.),

Rodwell TC, Kapasi AJ, Moore M, Milian-Suazo F, Harris B, Guerrero LP, Moser K, Strathdee SA, Garfein RS. (2010). Tracing the origins of Mycobacterium bovis tuberculosis in humans in the USA to cattle in Mexico using spoligotyping. Int J Infect Dis.;14 Suppl 3:e129-35

Roetzer A, Schuback S, Diel R, Gasau F, Ubben T, di Nauta A, Richter E, Rüsch-Gerdes S, Niemann S. 2011.Evaluation of Mycobacterium tuberculosis typing methods in a

# 2

# Measuring of DNA Damage by Quantitative PCR

Ayse Gul Mutlu

*Mehmet Akif Ersoy University, Arts and Sciences Faculty, Department of Biology, Burdur*
*Turkey*

## 1. Introduction

### 1.1 QPCR; principles and development

PCR is an in vitro method of nucleic acid synthesis by which a particular segment of DNA can be specifically replicated. It involves two oligonucleotide primer that flank the DNA fragment to be amplified and repeated cycles of heat denaturation of the DNA, annealing of the primers to their complementary sequences, and extension of the annealed primers with DNA polymerase. Successive cycles of amplification essentially double the amount of the target DNA synthesized in the previous cycle (1).

Recent advances of the fluorometric dyes allow the very sensitive and quick quantitation of DNA. Before the invention of fluorometric quantitative PCR (QPCR) method the researchers who measured a gene's amount, have used the different methods like competetive PCR, solid phase assays, HPLC, dot blot or immunoassay (2). Many of the applications of real-time Q-PCR include measuring mRNA expression levels, DNA copy number, transgene copy number and expression analysis, allelic discrimination, and measuring viral titers (3).

The detection of gene-spesific damage and repair has been studied in nuclear and mitochondrial DNA by the use of southern analysis. But this method requires knowledge of the restriction sites flanking the damaged site, the use of large quantities of DNA, and incision of DNA lesions with a spesific endonuclease (4). Govan and collegues has reported a new approach to measuring of DNA damage in 1990 (5). This PCR based quantitative technique has been improved by Kalinowski and collegues (6). Principle of this analysis is that lesions present in the DNA, block the progression of any thermostable polymerase on the template. So the DNA amplification decreases in the damaged template when compared to the undamaged DNA (4). QPCR is a suitable method for the measuring damage and repair in the subgene level functional units like promotor regions, exons and introns (7). Method also useful to determining DNA damage and repair that originated by the genotoxic agents and oxidative stress (8,9,10). The method capable of detect 1 lesion/$10^5$ nucleotides from as little as 5 ng of total genomic DNA (4).

DNA extraction, pre quantitation of DNA template, PCR amplification and quantitation of PCR products are crucial for success of the application (Figure 1). Quantity and quality of the DNA sample is important. We use mini column based kits for DNA extraction in our laboratory. These extraction kits and carefully pipetting, minimize the artificial DNA damages. Pico Green dsDNA quantitation kit is used for both template DNA quantitation and the analysis of PCR products as fluorometrically 485 nm excitation, 530 nm emission. Pico Green and SYBR green are substantially more sensitive for quantifying DNA concentrations than ethidium bromide and some other fluorimetric dyes (11). Initial DNA template quantity in the all PCR tubes must be the same. mtDNA damage is quantified by comparing the relative efficiency of amplification of long fragments of DNA and normalizing this to gene copy numbers by the amplification of smaller fragments, which have a statistically negligible likelihood of containing damaged bases. To calculate relative amplification, the long QPCR values are divided by the corresponding short QPCR results to account for potential copy number differences between samples. Decreased relative amplification is an indicator of the damaged DNA (4, 12).

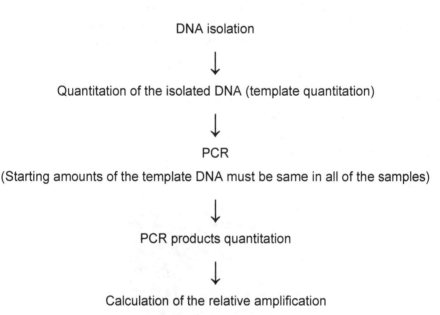

Fig. 1. Flowchart of the QPCR assay for measuring of DNA damage

## 2. Optimization of the assay; the crucial steps

Crucial step of the QPCR is PCR optimization. Thermal conditions, especially annealing temperature must be optimized. Extention temperature may be lower for long PCR amplifications. mtDNA amplification may needs some adjuvants. We use in our laboratory DMSO (%4) for improve the efficiency of the PCR reaction. Various authors recommend DMSO and glycerol to improve amplification efficiency (higher amount of product) and specificity (no unspecific products) of PCR, when used in concentrations varying between 5%-10% (vol/vol). However, in the multiplex reactions, these adjuvants give conflicting results. For example, 5% DMSO improve the amplification of some products and decrease the amount of others. There are similar results with 5% glycerol. Therefore, the usefulness of these adjuvants needs to be tested in each case. Also BSA may increase the efficiency of the PCR (13).

Hot start PCR improve specifity of PCR reaction. Hot start PCR is reported to minimize nontarget amplification and the formation of primer-dimer (14) .

Optimization might involve changes in nucleic acids preparation, in primer usage, in buffer usage and in cycling parameters. One of the recent developments in PCR optimization is to recognize the importance of eliminating some undesired hybridization events that often happen in the first cycle and can carry potentially devastating effects. Theoretically, if the amplification precedes with an efficiency of 100%, the amount of amplicons is doubling at each cycle. However, in most PCR procedures, the overall efficiency is less than 100% and a typical amplification runs with a constant efficiency of about 70-80% from the 15th cycle to the 30th cycle, depending on the amount of starting material. The increase in the amount of amplicons stays exponential only for a limited number of cycles, after which the amplification rate reaches a plateau. The factors that contribute to this plateau phenomenon include substrate saturation of enzyme, product strand reannealing, and incomplete product strand separation. In this latter phase, the quantitated amount of amplified product is no longer proportional to the starting amount of target molecules. Therefore, to make PCR suitable in quantitative settings, it is imperative that a balance be found between a constant efficiency and an exponential phase in the amplification process. This will ultimately depend on the number of cycles, on the amount of targets in the starting material, and on the system of detection and quantitation of the amplified product (15).

We run a 50% template control and a nontemplate control in PCR in our laboratory. 50% template control should given a 50% reduction of the amplification signal (values between 40%-60% reduction are acceptable). The nontemplate control would detect contamination with spurious DNA or PCR products (4).

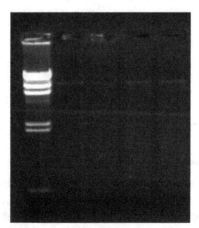

Fig. 2. PCR band of 10 kb mtDNA fragment of Mus musculus (Balb C). (Band 1: λ *Hind III* digest marker DNA)

## 3. Measuring of mtDNA damage on mice

We study mtDNA damage by QPCR method in different organisms like fruit flies, mice and snails in our laboratory. In our research that we used mice, oxidative mtDNA damage that created by cigarette smoke and protective effects of VitE and selenium was investigated (Figure 3).

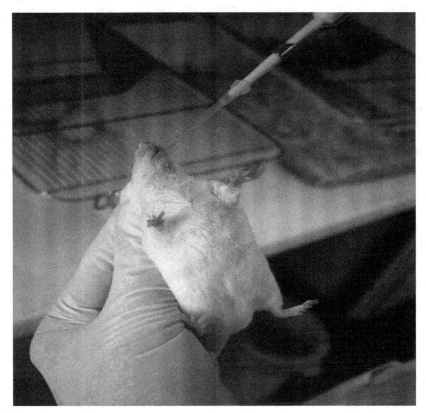

Fig. 3. Vitamin supplementation

DNA damage that is originated by cigarette smoke in various organs is declared by some research (16,17). Tobacco smoking contains many thousands of chemicals including a plethora of mutagens. Many carcinogens undergo metabolic activation in mammalian tissues to reactive intermediates that interact with and modify informational macromolecules, such as DNA with potentially mutagenic consequences (18). PAHs (Polycyclic aromatic hydrocarbons) cause irreversible DNA damage via covalent binding or oxidation (19). However genetic damage reflecting individual exposure and susceptibility to PAH may play a role in disease development (20). Tobacco smoke contains major classes of carcinogens that include PAHs, aromatic amines and tobaccospecific nitrosamines. In addition, toxic compounds such as formaldehyde, acetaldehyde, acrolein, short-lived radicals and reactive oxygen intermediates generated by redox cycling from catechol and hydroquinone and nitric oxide (NO) may also contribute to the toxic and carcinogenic effects of tobacco smoke. Direct DNAdamaging compounds that are present in cigarette smoke (CS) have previously been reported to include reactive oxygen intermediates, peroxynitrite, ethylating agents and unidentified compounds (21).

Many carcinogens in the cigarette smoke like PAHs, nitrosamine and cisplatin bind mitochondrial DNA (mtDNA) preferentially (22). The antioxidants are used frequently as food supplements may be effective to preventing cigarette smoke damage on mtDNA. Damages that are created by CS may be prevented by vitamin E (Vit E) and selenium (Se) which are powerful antioxidants.

Genomic DNA mini column kit (SIGMA) was used for total DNA isolation according to the technical bulletin. We used Pico Green dsDNA quantitation kit for both template DNA quantitation and the analysis of PCR products as fluorometrically 485 nm excitation, 530 nm emission (23). A crucial step of quantitative PCR is the concentration of the DNA sample. In fact, the accuracy of the assay relies on initial template quantity because all of the samples must have exactly the same amount of DNA. The Pico Green dye has not only proved efficient in regarding to template quantitation but also to PCR product analysis (10). Hot Start ready mix Taq (SIGMA) were used for PCR. In this mix, taq polymerase combines the performance enhancements of Taq antibody for hot start. When the temperature is raised above 70°C in the first denaturation step of the cycling process, the complex dissociates and the polymerase becomes fully active. DMSO as 4% of total volume and 20 ng of template total DNA were added into the each PCR tube. Mouse 117 bp Mouse 117 bp mtDNA fragment (small fragment) primers were:

13597 5'- CCC AGC TAC TAC CAT CAT TCA AGT- 3'

13688 5'- GAT GGT TTG GGA GAT TGG TTG ATG T- 3' (Table 1)

| |
|---|
| *Mus musculus* primers for long fragment (10085 bp): <br> 3278  5'- GCC AGC CTG ACC CAT AGC CAT AAT AT- 3' <br> 13337 5'- GAG AGA TTT TAT GGG TGT AAT GCG G- 3'  (4) |
| *Mus musculus* primers for small fragment (117 bp): <br> 13597 5'- CCC AGC TAC TAC CAT CAT TCA AGT- 3' <br> 13688 5'- GAT GGT TTG GGA GAT TGG TTG ATG T- 3' (4) |
| *Drosophila* primers for long fragment (10629 bp): <br> 1880    5'- ATGGTGGAGCTTCAGTTGATTT - 3' <br> 12487  5'- CAACCTTTTTGTGATGCGATTA - 3' |
| *Drosophila* primers for small fragment (100 bp): <br> 11426   5'- TAAGAAAATTCCGAGGGATTCA - 3' <br> 11504   5'- GGTCGAGCTCCAATTCAAGTTA - 3' |

Table 1. QPCR primers for measuring of mtDNA damage in *Mus musculus* and *Drosophila melanogaster*

Thermal conditions for *Mus musculus* long fragment (10085 bp):

75°C for 2 min

95°C for 1 min

**94°C for 15 sec**

**59°C for 30 sec   → 21 cycles**

**65 °C for 11 min**

72°C for 10 min

---

Thermal conditions for *Mus musculus* small fragment (117 bp):

75 °C for 2 min

95 °C for 15 sec

**94°C for 30 sec**

**50°C for 45 sec   → 19 cycles**

**72 °C for 45 sec.**

72°C for 10 min

---

Thermal conditions for *Drosophila* long fragment (10629 bp):

75°C for 1 min

95°C for 1 min

**94°C for 15 sec**

**52°C for 45 sec   → 21 cycles**

**65 °C for 11 min**

68°C for 10 min.

---

Thermal conditions for *Drosophila* small fragment (100 bp):

75 °C for 2 min

95 °C for 15 sec

**94°C for 30 sec**

**55°C for 45 sec   → 21 cycles**

**72 °C for 45 sec**

72°C for 10 min

Table 2. Thermal conditions for QPCR in *Mus musculus* and *Drosophila melanogaster*

Mouse 10 kb mtDNA fragment (Figure 2) primers were:

3278 5'- GCC AGC CTG ACC CAT AGC CAT AAT AT- 3'

13337 5'- GAG AGA TTT TAT GGG TGT AAT GCG G- 3'

(4).

For long fragment PCR amplification, DNA was denatured initially at 75°C for 2 min and 95°C for 1 min, and then the reaction underwent 21 PCR cycles of 94°C for 15 sec, 59°C for 30 sec, and 65 °C for 11 min. Final extension was allowed to proceed at 72°C for 10 min (Table 2). For small fragment PCR amplification, DNA was denatured initially at 75 °C for 2 min and 95 °C for 15 sec, and then the reaction underwent 19 PCR cycles of 94°C for 30 sec, 50°C for 45 sec, and 72 °C for 45 sec. Final extension was allowed to proceed at 72°C for 10 min (23).

We were always run a 50% template control and a nontemplate control in PCR. To calculate relative amplification, the long QPCR values were divided by the corresponding short QPCR results to account for potential copy number differences between samples (mtDNA/total DNA value may be different in 20 ng template total DNA of each PCR tube) (3,4,10,23). The copy number results not indicate the damage.

We detected mtDNA damage in the mouse heart succesfully. According to these relative amplification results "cigarette smoke" application group was significantly different from all other groups. mtDNA damage was significantly higher in the cigarette smoke group than the other groups. However "Cigarette Smoke+Vitamin E+Selenium" group had lowest mean damage (23,24).

## 4. Measuring of oxidative mtDNA damage and copy number on Drosophila

The free radical theory of aging postulates that aging changes are caused by free radical reactions. Aging is the progressive accumulation of changes with time that are responsible for the ever-increasing likelihood of disease and death. These irreversible changes are attributed to the aging process. This process is now the major cause of death in the developed countries. The aging process may be due to free radical reactions (25). The free radical theory of aging posits that the accumulation of macromolecular damage induced by toxic reactive oxygen species (ROS) plays a central role in the aging process. The mitochondria are the principal generator of ROS during the conversion of molecular oxygen to energy production where approximately 0.4% to 4% of the molecular oxygen metabolized by the mitochondrial electron transport chain is converted to ROS (26). Cellular damage caused by radicals may induce cancer, neurodegeneration and autoimmun disease (27). Toxic materials may produce ROS and generate oxidative damage on mitochondrial DNA (mtDNA) (23). mtDNA damages may trigger mitochondrial dysfunction (28). Damage to mtDNA could be potentially more important than deletions in nDNA, because the entire mitochondrial genome codes for genes that are expressed while nDNA contains a large amount of non-transcribed sequences. Also, mtDNA, unlike nDNA, is continuously replicated, even in terminally differentiated cells, such as neurons and cardiomyocytes; hence, somatic mtDNA damage potentially causes more adverse effects on cellular functions than does somatic nDNA damage (29).

Cereals naturally contain a wide variety of polyphenols such as the hydroxycinnamic acids, ferulic, vanillic, and *p*-coumaric acids which show a strong antioxidant power and may help to protect from oxidative stress and, therefore, can decrease the risk of contracting many diseases. Flavonoids are present in small quantities, even though their numerous biological effects and their implications for inflammation and chronic diseases have been widely described. The mechanisms of action of polyphenols go beyond the modulation of oxidative stress-related pathways (30).

Wheat is an important component of the human diet. But the distribution of phytochemicals (total phenolics, flavonoids, ferulic acid, and carotenoids) and hydrophilic and lipophilic antioxidant activity in milled fractions (endosperm and bran/germ) are different each other. Different milled fractions of wheat have different profiles of both hydrophilic and lipophilic phytochemicals. Total phenolic content of bran/germ fractions is 15–18-fold higher than that of endosperm fractions. Hydrophilic antioxidant activity of bran/germ samples is 13–27-fold higher than that of the respective endosperm samples. Similarly, lipophilic antioxidant activity is 28–89-fold higher in the bran/germ fractions. In whole-wheat flour, the bran/germ fraction contribute 83% of the total phenolic content, 79% of the total flavonoid content, 51% of the total lutein, 78% of the total zeaxanthin, 42% of the total β-cryptoxanthin, 85% of the total hydrophilic antioxidant activity, and 94% of the total lipophilic antioxidant activity (31).

Aim of our study was investigate the effects of a wheat germ rich diet on oxidative mtDNA damage, mtDNA copy number and antioxidant enzyme activities in the aging process of *Drosophila* (32).

Genomic DNA kits (invitrogen) were used for total DNA isolation according to the technical bulletin. İnvitrogen (Molecular Probes) Pico Green dsDNA quantitation dye and QUBIT 2.0 fluorometer were used for both template DNA quantitation and the analysis of PCR products as fluorometrically (Figure 4). DMSO as 4% of total volume and 5 ng of template total DNA were added into the each PCR tube.

Primers for Drosophila mtDNA 100bp fragment were designed as;

11426   5′- TAAGAAAATTCCGAGGGATTCA - 3′

11525   5′- GGTCGAGCTCCAATTCAAGTTA - 3′

Primers for Drosophila mtDNA 10629 bp fragment were designed as;

1880    5′- ATGGTGGAGCTTCAGTTGATTT - 3′

12508   5′- CAACCTTTTTGTGATGCGATTA - 3′ (Table 1)

For long fragment PCR amplification, DNA was denatured initially at 75°C for 1 min and 95°C for 1 min, and then the reaction underwent 21 PCR cycles of 94°C for 15 sec, 52°C for 45 sec, and 65 °C for 11 min. Final extension was allowed to proceed at 68°C for 10 min (Table 2).

For small fragment PCR amplification, DNA was denatured initially at 75 °C for 2 min and 95 °C for 15 sec, and then the reaction underwent 21 PCR cycles of 94°C for 30 sec, 55°C for 45 sec, and 72 °C for 45 sec. Final extension was allowed to proceed at 72°C for 10 min.

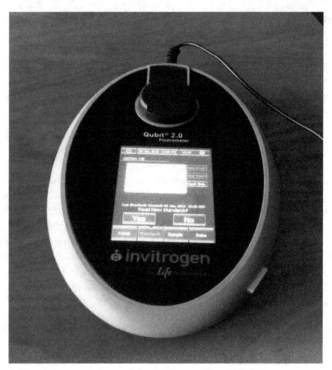

Fig. 4. QUBIT 2.0 fluorometer were used for both template DNA quantitation and the analysis of PCR products as fluorometrically

## 5. Conclusions

QPCR is a suitable method for the measuring damage and repair in the subgene level functional units like promotor regions, exons and introns (7). Recent advances of the fluorometric dyes allow the very sensitive and quick quantitation of DNA. Before the invention of fluorometric quantitative PCR (QPCR) method, the researchers who measured a gene's amount, have used the different methods like competetive PCR, solid phase assays, HPLC, dot blot or immunoassay (2). Many of the applications of real-time Q-PCR include measuring mRNA expression levels, DNA copy number, transgene copy number and expression analysis, allelic discrimination, and measuring viral titers (3). Method also useful to determining DNA damage and repair that originated by the genotoxic agents and oxidative stress (8,9,10). Crucial step of the QPCR is PCR optimization. Thermal conditions, especially annealing temperature must be optimized. Important points of the optimization:

1. Determination of annealing temperature
2. Optimization of he extention temperature (Extention temperature may be lower for long PCR amplifications)
3. Adjuvants (if necessary)
4. Hot start PCR (minimize nontarget amplification and the formation of primer-dimer)
5. Determination of cycling number

6. Running of %50 template and nontemplate controls in PCR (50% template control should given a 50% reduction of the amplification signal -values between 40%-60% reduction are acceptable-. The nontemplate control would detect contamination with spurious DNA or PCR products)

We detected mtDNA damage that originated by the genotoxic agents, oxidative stress and age, above mentioned conditions in our various studies (23,24,32). Also, QPCR method is suitable for the nutritional studies and some cancer researches.

## 6. References

[1] Saiki RK, Amplification of genomic DNA. PCR Protocols: a guide to methods and applications (eds. Innis, Gelfand, Sninsky, White). Academic Press, California. 1990.

[2] Wong A, Cortopassi G, Reproducible QPCR of mitochondrial and nuclear DNA copy number using the LightCycler. Mitochondrial DNA methods and Protocols (eds.Copeland). pp.129-149, Humana Press Inc, Totawa, NJ. 2002.

[3] Ginzinger DG, 2002. Gene quantification using real-time quantitative PCR: An emerging technology hits the mainstream. Experimental Hematology, 30(6): 503-512.

[4] Santos JH, Mandavilli BS, Van Houten B, Measuring oxidative mtDNA damage and repair using QPCR. Mitochondrial DNA methods and Protocols (eds.Copeland). pp.159-176, Humana Press Inc, Totawa, NJ. 2002.

[5] Govan HL, Valles-Ayoub Y, Braun J, 1990. Fine-mapping of DNA damage and repair in specific genomic segments. Nucleic Acids Research, 18 (13):3823-3830.

[6] Kalinowski DP, Illenye S, Van Houten B, 1992. Analysis of DNA damage and repair in murine leukemia L1210 cells using a QPCR assay. Nucleic Acids Research, 20(13):3485-3494.

[7] Grimaldi KA, Bingham JP, Souhami RL, Hartley JA, 1994. DNA damage by anticancer reagents and its repair: mapping in cells in the subgene level with QPCR reaction. Anal Biochem, 222(1):236-242.

[8] Van Houten B, Cheng S, Chen Y, 2000. Measuring gene spesific nucleotide excision repair in human cells using quantitative amplification of long targets from nanogram quantities of DNA. Mutat Res, 25; 460(2):81-94.

[9] Ayala-Torres S, Chen Y, Suoblada T, Rosenblatt J, Van Houten B, 2000. Analysis of gene spesific DNA damage and repair using QPCR. Methods, 22: 135-147.

[10] Yakes FM, Van Houten B, 1997. Mitochondrial DNA damage is more extensive and persists longer than nuclear DNA damage in human cells following oxidative stres. Proc Natl Acad Sci USA, 94: 514-519.

[11] Rengarajan K, Cristol SM, Mehta M, Nickerson JM, 2002. Quantifying DNA concentrations using fluorimetry: A comparison of fluorophores. Molecular Vision, 8: 416-421.

[12] Venkatraman A, Landar A, Davis AJ, Chamlee L, Sandersoni T, Kim H, Page G, Pompilius M, Ballinger S, Darley-UsmarV, Bailey SM, 2004. Modification of the mitochondrial proteome in response to the stres of ethanol-dependent hepatoxicity. J Biol Chem, 279: 22092-22101.

[13] O. Henegariu, N.A. Heerema, S.R. Dlouhy, G.H. Vance and P.H. Vogt, 1997. Multiplex PCR: Critical Parameters and Step-by-Step Protocol. BioTechniques, 23: 504-511.

[14] Erlich HA, Gelfand D, Sninsky JJ, 1991. Recent Advances in the Polymerase Chain Reaction. Science, 252: 1643-1651.

[15] Ferre F, 1992. Quantitative or semi-quantitative PCR: reality versus myth. Genome Research, 2:1-9.

[16] Izzotti A, Balanksy RM, Blagoeva PM, Mircheva Z, Tulimiero L, Cartiglia C, De Flora S, 1998. DNA alterations in rat organs after chronic exposure to cigarette smoke and/or ethanol digestion. FASEB J, 12: 753-758.

[17] Izzotti A, Bagnasco M, D'Agostini F, Cartiglia C, Lubet RA, Kelloff G, De Flora S, 1999. Formation and persistence of nucleotide alterations in rats exposed whole-body to environmental cigarette smoke. Carcinogenesis, 20: 1499-1506.

[18] Phillips DH, 2002. Smoking related DNA and protein adducts in human tissues. Carcinogenesis, 23: 1979- 2004.

[19] Gelboin HV, 1980. Benzo α pyrene metabolism, activation and carcinogenesis: role and regulation of mixed-function oxidases and related enzymes. Physiol Rev, 60:1107–66.

[20] Rundle A, Tang D, Hibshoosh H, Estabrook A, Schnabel F, Cao W, Grumet S, Perera FP, 2000. The relationship between genetic damage from polycyclic aromatic hydrocarbons in breast tissue and breast cancer. Carcinogenesis, 21: 1281-1289.

[21] Yang Q, Hergenhahn M, Weninger A, Bartsch H, 1999. Cigarette smoke induces direct DNA damage in the human B-lymphoid cell line Raji. Carcinogenesis, 20: 1769-1775.

[22] Sawyer DE, Van Houten B. (1999) Repair of DNA damage in mitochondria. Mutat Res, 434: 161- 176.

[23] Mutlu AG, Fiskin K, 2009. Can Vitamin E and Selenium Prevent Cigarette Smoke-Derived Oxidative mtDNA Damage? Turkish Journal of Biochemistry, 34 (3); 167-172.

[24] Mutlu AG., Fiskin K, 2009. Oxidative mtDNA damage in the heart tissue of cigarette smoke exposed mice and protective effects of Vitamın E and Selenium. IUBMB Life. 61: 328-329. (III. International Congress of Molecular Medicine, İstanbul, 5-8 May 2009).

[25] Harman D, 2006. Free Radicals in Aging. Moll Cell Biochem, 84: 155-61.

[26] Lim H, Bodmer R and Perrin L, 2006. Drosophila aging 2005-2006. Exp Gerontol, 41: 1213-1216.

[27] Rodriguez C, Mayo JC, Sainz RM, Antolin I, Herrera F, Martin V and Reiter RJ, 2004. Regulation of antioxidant enzymes: a significant role for melatonin. J Pineal Res, 36: 1-9.

[28] Lesnefsky EJ, Moghaddas S, Tandler B, Kerner J and Hoppel CL, 2001. Mitochondrial dysfunction in cardiac disease: ischemia –reperfusion, aging and heart failure. J Mol Cell Cardiol, 33: 1065-1089.

[29] Liang F-Q and Godley BF, 2003. Oxidative stres induced mtDNA damage in human retinal pigment epithelial cells: a possible mechanism for RPE aging and ge-related macular degeneration. Experimental Eye Research, 76: 397-403.

[30] Alvarez P, Alvarado C, Puerto M, Schlumberger A, Jimenez L and De la Fuente M, 2006. Improvement of leucocyte functions in prematurely aging mice after five weeks of diet supplementation with polyphenol-rich cereals. Nutrition, 22: 913-921.

[31] Adom KK, Sorrells ME and Liu RH, 2005. Phytochemicals and antioxidant activity of milled fractions of different wheat varieties. J Agric Food Chem, 53: 2297-2306.

[32] Mutlu AG, 2011. Bugday embriyosunca zengin bir diyetin, Drosophila'nın yaslanma surecinde, oksidatif mtDNA hasarı, mtDNA kopya sayısı, ve antioksidan enzim aktiviteleri üzerine etkileri. Türkiye Klinikleri Tıp Bilimleri Dergisi, 31(6): 132.

# 3

# Detection of *Apple Chlorotic Leaf Spot Virus* in Tissues of Pear Using *In Situ* RT-PCR and Primed *In Situ* Labeling

Na Liu, Jianxin Niu* and Ying Zhao
*Department of Horticultural, Agricultural College of Shihezi University, Shihezi*
*People's Republic of China*

## 1. Introduction

*Apple chlorotic leaf spot virus* (ACLSV) is the type member of the Trichovirus genus, the family *Flexiviridae* (Martelli et al., 1994; Adams et al., 2004) and is known to infect most pome and stone fruit tree species, including apple, peach, pear, plum, almond, cherry and apricot (Lister, 1970; Németh, 1986). ACLSV has a worldwide distribution and induces a large variety of symptoms in sensitive fruit trees (Németh, 1986; Dunez & Delbos, 1988; Desvignes & Boyé, 1989). However, In Japan, this virus is one of the causative agents of topworking disease and induces lethal decline in apple trees grown on Maruba kaido (Malus prunifolia var. ringo) rootstocks (Yanase, 1974). Other severe symptoms of stone fruit trees in Europe caused by ACLSV including bark split and pseudopox in plum, bark split in cherry, pseudopox and graft incompatibility in apricot and ring pattern mosaic in pear (Dunez et al., 1972; Desvignes & Boyé, 1989; Cieślińska et al., 1995; Jelkmann & Kunze, 1995). ACLSV has very flexuous filamentous particles, approximately 640 to 760 nm in length and consisting of a single-stranded positive-sense RNA with Mr of $2.48 \times 10^6$ and multiple copies of a 22 kDa coat protein (CP) (Yoshikawa & Takahashi, 1988).

*In situ* detection techniques allow specific nucleic acid sequences to be exposed in morphologically preserved tissue sections. In combination with immunocytochemistry, *in situ* detection can relate microscopic topological information to gene activity at the transcript or protein levels in specific tissues. In certain cases, they also can provide increased specificity and more rapid analyses. *In situ* reverse transcription polymerase chain reaction (RT-PCR) is a molecular biological-cytological method. *In situ* RT-PCR combined the sensitiveness of PCR amplification with spatial localization of products to monitor the appearance of specific transcripts in the tissue sections. Therefore, *in situ* RT-PCR defined a powerful tool for the low abundance transcript detection (Pesquet et al., 2004). Hasse et al. (1990) first reported the *in situ* PCR technology, which combined the strong points of PCR and *in situ* hybridization. It was widely used for all kinds of disease and genetic studies in human and animal (Gressens & Martin, 1994; Staskus et al., 1991; Nuovo et al., 1991; Bagasra et al., 1992; Cohen, 1996; Chen & Fuggle, 1993; Höfler et al., 1995). The first application of *in situ* RT-PCR for the plant tissue was reported by Woo et al. (1995). Most recently, this

---

* Corresponding Author

method had not been used to a large extent in plants (Greer et al., 1991; Johansen, 1997; Matsuda et al., 1997).

The primed *in situ* labeling (PRINS) procedure is a fast and efficient alternative to conventional fluorescence *in situ* hybridization for nucleic acid detection. According to the PRINS method, laboratory-synthesized oligonucleotide probes are used instead of cloned DNA for the *in situ* localization of individual genes. The PRINS primers are annealed to complementary target sequences on tissues and are extended in the presence of labeled nucleotides (Koch et al., 1995) utilizing *Taq* DNA polymerase. Since its introduction, the PRINS protocol has been continuously optimized, and numerous applications have been developed (Thomas et al., 2001; Yan et al., 2001; Xu et al., 2002; Tharapel & Wachtel, 2006a, 2006b; Wachtel & Tharapel, 2006; Kaczmarek et al., 2007). The technique has thus proved to be a useful tool for *in situ* screening, and has become a simple and efficient complement to conventional and molecular cytogenetic methods.

In this paper, we optimized the *in situ* RT-PCR and PRINS method for increased sensitivity to localize the virus in plant tissues with ACLSV. Based on this research, through observing distribution of amplified cDNA in tissues, we can analysis the virus infection. In this way, it can provide a new approach to detection virus in fruit trees, as well as investigate the formation, distribution and transformation of virus and produce innocuity fruit trees.

## 2. Materials

### 2.1 Virus sources

Leaves were collected from Korla pear in Shayidong commercial orchard of Korla, Xinjiang, China. Virus-free healthy leaves were used as negative controls.

### 2.2 Reagents and enzymes

*Taq* DNA Polymerase, dNTPs, dATP, dGTP, dCTP, dTTP, PMD19-T were all purchased from TakaRa (China); M-MLV Reverse Transcriptase, T4 DNA ligase were from Fermentas (USA); TIANprep Mini Plasmid Kit and TIANgel Midi purification Kit were from TIANGEN (China); SuperScript II RNase H-Reverse Transcriptase were from Invitrogen (EU); Proteinase K were from Merk (Germany); Digoxigenin-11- dUTP, alkaline phosphatase labeled anti-digoxin, anti-digoxin- fluorescence, Ribonuclease inhibitor, DNaseI were purchased from ROCH (USA); Nitro blue tetrazolium chloride (NBT)/5-bromo-4-chloro- 3-indolylphosphate (BCIP) were purchased from Shanghai Sangon (China); others were all analysis purity made in China. *E. coli* DH5α as preserved strains were stored at Biotechnology Laboratory of Horticultural Department, Agriculture College, Shihezi University, China.

### 2.3 Primer design

The sequences were amplified by *in situ* RT-PCR reaction with specific primers, which were designed according to the cDNA sequence of ACLSV (Sato et al., 1993). Primer sequences are as follows: forward primer (P3) 5'-GGCAACCCTGGAACAGA-3' and the reverse primer (P4) 5'-CAGACCCTTATTGAAG TCGAA-3'.

The sequences were amplified by PRINS reaction with specific primers, which were designed according to the cDNA sequence of ACLSV from GenBank D14996 (Table 1). A

Blast search of the primer sequences showed that they were specific for their intended targets.

| Primer | Primer Sequence (5'-3') | Annealing Temp (°C) |
|--------|-------------------------|---------------------|
| acls Pa 1 | CTTTACGAGCCCATTTCTTGCC | 61.5 |
| acls Ps 1 | GAACATAGCGATACAGGGGACC | 60.3 |
| acls Pa 2 | TGCCTCACACACTTGGCGGAG | 60.6 |
| acls Ps 2 | CGATACAGGGGACCTCGGAAC | 61.5 |
| acls Pa 3 | GCCTTTACGAGCCCATTTCTTG | 59.5 |
| acls Ps 3 | AGGGGACCTCGGAACAAACAG | 60.5 |
| acls Pa 4 | GTACAAAAGAGGTTTGTGAAG | 54.2 |
| acls Ps 4 | GTGCTGGTGGAGGTGAAATC | 57.4 |
| acls Pa 5 | CAATCTGAAGGAGGTAGTCGGT | 56.4 |
| acls Ps 5 | TTCAGGCGTAGTAGAAAAGAGG | 57.7 |

Table 1. Oligonucleotide primers used to PRINS

## 3. Methods

### 3.1 Total RNA extraction and RT-PCR

Total RNAs were extracted from phloem infected by ACLSV. The 200 mg fresh Pear phloem tissue were grinded in liquide nitrogen for a fine powder and transferred to a 1.5 mL eppendorf tube which has added 800 µL extraction buffer (50 mmol·L$^{-1}$ Tris-Cl pH 8.0, 140 mmol·L$^{-1}$ NaCl, 10 mmol·L$^{-1}$ EDTA, 4% SDS, 3% PVP, 15% ethanol, 5% β-mercaptoethanol), well mixed by invertion of the tube. Added 500 µL Tris-saturated phenol (pH 8.0): chloroform: isoamyl alcohol (25: 24: 1) to the tube, sepaeated by centrifugation at 12 000 rpm for 15 min at 4°C. Transferred the supernatant by hand-suction to a fresh tube and mixed with an equal volume of Tris-saturated phenol (pH 8.0): chloroform: isoamyl alcohol (25: 24: 1), followed by centrifugation at 12 000 rpm at 4 °C for 15 min. The supernatant was transferred to a fresh tube and mixed with an equal volume of chloroform: isoamyl alcohol (24: 1) and then centrifugation at 12 000 rpm at 4°C for 10 min. Transferred the supernatant to a fresh tube and added 2.0 volumes of LiCl. Precipitated at –20°C for 2-3 h. RNA was separated by centrifugation at 12 000 rpm for 15 min at 4°C. Removed the supernatant by hand-suction, washed the pellet two times by 70% ethanol, air-dry at room temperature. Suspended the pellet in 20-30 µL of TE solution or DEPC-treated sterile water and analysed it immediately by electrophoresis or stored at – 20°C.

The reverse transcription mixture contained 1.0 µL specific reverse primer and 5.0 µL of total RNA and 9.5 µL of ddH$_2$O. The mixture was kept at 70°C for 5 min, and then immediately transferred to ice for 5 min. Then 2.5 µL of dNTPs (10 mM each), 5.0 µL of 5×M-MLV buffer, 1.0 µL of RNasin ribonuclease inhibitor (40 U µL$^{-1}$), 1.0 µL of M-MLV reverse transcriptase (200 U µL$^{-1}$) and made the total volume of 25.0 µL. The mixture was incubated at 42°C for 1 h.

PCR reaction volumes were 20.0 μL, and contained 2.0 μL of 10×PCR buffer, 0.5 μL of dNTPs (each 10 mM), 2.0 μL of primers, 2.0 μL of cDNA, 0.2 μL (5U μL⁻¹) *Tap* DNA polymerase and 13.3 μL of ddH$_2$O. PCR was carried out with an initial denaturation of 4 min at 94°C, followed by 35 cycles of 30s, 94°C; 30s, 55°C; 1 min, 72°C; and then by a final elongation step of 7 min at 72°C.

### 3.2 Cloning and sequencing

The amplified PCR products were gel purified and extracted using TIANgel Midi Purification Kit (TIANGEN, China). The purified DNA fragments were ligated into the PMD19-T vector (TaKaRa Biotechnology, China) following the manufacturer's instruction, and used to transform *E. coli* DH5α. The positive clones were confirmed by PCR and restriction enzyme digestion before sequencing. Two clones from independent PCR reactions were sequenced from both directions.

### 3.3 Tissue embedding and preparation of slide

1. Slide disposal: After rinsed, ultrasonic cleaned and high temperature baked, the slide must be pre-prepared with poly-L-lysine for 5 min, and then incubated it at 26°C overnight, sealed and stored at room temperature for use within 10 d.
2. Tissues fixation: Leaves were cut into small pieces (3×2 mm) and rinsed the tissues in 4% paraformaldehyde immediately for 1h at room temperature with gentle shaking.
3. Dehydration: Washed the tissues in PBS buffer two times (5 min each), immersed the tissues in series of concentration of ethanol (50%, 70%, 85%, 95% and 100%) for 1h, respectively, at room temperature.
4. Transparences: Put the tissues into pure alcohol: xylene (1: 1) and pure xylene for 1 h, respectively, at room temperature.
5. Low-temperature wax infiltration: Put the tissues into the container which contained transparence and paraffin, covered the container with lid, and incubated at 38°C overnight.
6. High-temperature wax infiltration: Removed the lid, and put the container into incubator at 58°C, and then changed the pure paraffin three times for 2 h each.
7. Paraffin-embedding: Pour melted paraffin wax to pre-folded carton for embedding.
8. Sectioning: Tissue sections (2-16 μm) were obtained by a conventional rotary microtome. If very thin sections were required, a retracting rotary microtome should be used to avoid the compression of the tissue block by the up-stroke of the knife and sections should be mounted onto poly-L-lysine-coated pre-prepared slides.
9. Stretched section: Wax sections needed to be stretched before adhesion to the glass slide. Sections were lifted onto a layer of de-gassed water on a slide held on a warmed flat plate (45°C). Once the sections was stretched, drained away the excess water and left the slide into incubator at 40°C, overnight, the section has dried onto the slide, stored at -20°C.

### 3.4 Pretreatment of slides

1. De-waxed: Removed the slides from the refrigerator, put the slide into the oven incubated for 1-3 h, at 60°C in order to melt paraffin. Rinsed the slide in xylene for 5 min

and transferred to ethanol for 5 min, repeated more times until the paraffin was completely removed, then left the slide at room temperature for air-dry.

2. Proteinase K treatment: Added 1 µg·mL$^{-1}$ Proteinase K digested 10-45 min at 37°C, stopped reaction by washings for 5 min in PBS buffer and transferred to DEPC-treated sterile water for 5 min at room temperature, then air-dry.

3. DNaseI treatment: For each slide, 4.0 µL 10×DNase I buffer, 4.0 µL DNase I (10 U µL$^{-1}$), 1.0 µL Ribonuclease inhibitor (40 U µL$^{-1}$) and DEPC water added to 20.0 µL in a 0.5 mL microtube. Applied the reaction solution onto the slide and put it into humidified chamber and incubated at 37°C overnight.

4. Wash the slide two times in DEPC-treated sterile water for 5 min each and in alcohol for 5 min at room temperature.

## 3.5 *In situ* reverse transcription reaction

For each slide, 4.0 µL 5×Frist-Strand Buffer (MgCl$^{2+}$ 15 mM), 2.0 µL dNTPs (10 mM each), 1.0 µL RNasin (40 U µL$^{-1}$), 1.0 µL Antisense primer (20 µM), 2.0 µL DDT (0.1 M), 1.0 µL SuperScript II RT (200 U µL$^{-1}$), and DEPC water added to 20.0 µL in a 0.5 mL microtube. Applied the reaction solution onto the slide and put it into a humidified chamber and incubated at 42°C for 1 h, then inactived at 92°C for 1 min. Washed the slide two times for 5 min each in distilled water at room temperature.

## 3.6 *In situ* RT-PCR detection

### 3.6.1 *In situ* RT-PCR reaction

The reaction was consisted of 2.5 µL 10 × PCR buffer (Mg$^{2+}$ free), 0.5 µL dNTP (10 mmol µL$^{-1}$), 1.0 µL each primer (20 pmol µL-1), 2.5 µL Dig-11-dUTP (1 nmol µL$^{-1}$), 1.0 µL *Taq* DNA polymerase (2.5 U µL$^{-1}$) and distilled water to 25.0 µL. Mounted the slide with genic frame, added the reaction solution, and covered the slide with a cover slip ,then put the slide on the flate bloke of the thermocycler. Cycling parameters consisted of 94°C for 3 min, 94°C for 2 min and 35 cycles of a two-step PCR with an annealing temperature of 56°C for 1 min. Removed the cover slip and inactivated at 94°C for 2 min. Washed the slid two times for 10 min each in washing buffer with gentle shaking. Several slides were used as negative controls for each *in situ* RT-PCR experiment. One slide was healthy plant, the other slides were amplified without primers, *Taq* DNA polymerase, or RT step.

### 3.6.2 Immunoenzymatic detection

1. Mounted the slide with 100 µL blocking buffer (100 mmol·L$^{-1}$ Tris-HCl, pH 7.5, 150 mmol·L$^{-1}$ NaCl, and 3% BSA). Incubated the slide in a humidified chamber at 37°C for 30min. Drained the blocking buffer from the slide.

2. Added anti-Dig-alkaline phosphatase (1: 100 in blocking buffer), and incubated the slide in a humidified chamber for 30 min at room temperature.

3. Stopped the reaction by rinsing the slide with washing buffer (100 mmol·L$^{-1}$ Tris-HCl, pH 7.5, 150 mmol·L$^{-1}$ NaCl) two times for 10 min each at room temperature with gentle shaking.

4. Developed the color reaction by adding 100 µL of NBT/BCIP solution to the slide and incubated the slide in a humidified chamber for 60 min in the dark at room temperature. Then rinsed the slide with water to stop the reaction.

5. Rinsed the slide in series of concentration of ethanol, 50%, 70%, 85%, 95%, and 100% for 2 min, respectively, at room temperature for dehydration.
6. Put the slide into pure xylene for 3 min for transparent.
7. Covered the section with the cover slip using mounting solution, air-dry. Then the sections were ready for data recording, which could view under bright field microscopy through stained with Alcian Blue.

## 3.7 PRINS detection

### 3.7.1 PRINS reaction

1. Immersed slides in 0.02 N HCl for 20 min.
2. Denature the samples by immersing them in 70% formamide/2×SSC, at 72°C for 2 min.
3. Dehydrate the slides in a series (70%, 90%, and 100%) of ice-cold ethanol washes (4°C) before allowing them to air-dry.
4. Prepare reaction mixture in a final volume of 25.0 μL consisted of specific primers (20 μM) 10.0 μL, 0.1% BSA 2.5 μL, 0.2 mM dNTPs 2.5 μL (each), 0.02 mM dTTP 1.0 μL, 0.02 mM Dig-11-dUTP 3.0 μL, *Taq* buffer 2.5 μL, *Taq* DNA polymerase (2.5 U μL⁻¹) 1.0 μL and distilled water to 25.0 μL. Kept the mix on ice during preparation and until application to the slide.
5. Reaction mixture incubated at annealing temperature and incubated the denatured the slide for 7 min at annealing temperature. Applied the reaction mixture and covered the working area of the slide completely with a 22 × 22 cover slip on the denatured the slide, and then transferred to the heating block of the thermal.
6. Set up the PRINS program and start the reaction. The program was carried out on a programmable thermal cycler equipped with a flat plate for slides. The program consisted of one cycle of 9 min at annealing temperature with an additional 30 min at 72°C for extension.
7. After extension, the slide was removed from cycler, the cover slip was removed, and the slide washed in NE solution (500 mM NaCl, 50 mM EDTA, pH 8.0) at 72°C for 5 min, and transferred the slide to 4×SSC/0.2% Tween-20 at 50°C for 5 min to stop the reaction.

### 3.7.2 Visualization of PRINS products

1. For each slide, added 10 μg mL⁻¹ avidin-Rhodamine and 20 μg mL⁻¹ anti-digoxigenin-FITC.
2. Placed slides in a humidified chamber for 30 min at room temperature, worked in the dark as much as possible to avoid fluorescence bleaching.
3. The slide was rinsed in preheated solutions (1×PBS/0.2%Tween-20, 37°C; 0.5×PBS/0.2% Tween-20, 37°C; 0.2×PBS/0.2%Tween-20, 37°C) for 5min, respectively, air-dried.
4. Mounted the slide with 3μg mL⁻¹ of DAPI/antifade solution under a 22×22 coverslip counterstained for 10min, in dark.
5. Let the excess mounting medium dry. Approximately 1 h, permanently seal the slide with nail polish. Slide can be maintained at 4°C until scored.

### 3.7.3 Signal detection and image analysis

Olympus BX51 fluorescence microscope system was adopted for this process. This system contained Olympus UPlanFl 100×/1.30 Oil ∞/0.17 C1field lens, pass band filter with

DAPI/FITC/Rhodamine, AxioCam Camera module and Video Test-FISH 4.0 image analysis system.

## 4. Results

### 4.1 Detection ACLSV by RT-PCR

Total RNA were extract from the phloem of pear which were infected with ACLSV, first strand cDNA synthesis was obtained by reverse transcription using specific primer and 358 bp fragment was amplified by P3/P4 primers as shown in Figure. 1. The purified DNA fragments were ligated into the PMD19-T vector and transformed into *E. coli* DH5α. The positive clones were confirmed by PCR and restriction enzyme digestion before sequencing.

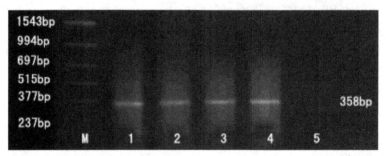

Fig. 1. The productions of RT-PCR of ACLSV
M: Marker; 1-4: productions; 5: negative control

### 4.2 Detection the reliability of alkaline phosphatase chromogenic system

The slide were digested by 1μg·mL⁻¹ Proteinase K for 20 min at 37°C, and incubated at 37°C overnight with DNase I. Washed the slide two times for 10 min each in PBS buffer. Mounted the slide with blocking buffer and incubated at 37°C for 30min. Added anti-Dig-alkaline phosphatase (1:100 in blocking buffer) and incubated the slide in a moist chamber for 60 min at room temperature, then washed the slide two times for 10 min each in PBS buffer at room temperature with gentle shaking. Added NBT/BCIP solution to the slide and incubated the slide in a humidified chamber for 60 min in the dark at room temperature. The result showed that sections were not stained.

### 4.3 The effect of treatment with proteinase K

After treated with Proteinase K treatment for 10 min or 15 min, the organization performed a piece of blue, which indicated that Proteinase K digested inadequately. Morphology was fuzzy when digested for 30 min or 40 min, illustrating excessive digestion. Proteinase K treatment 20 min was more moderate.

### 4.4 The effect of RT-component concentration

The results showed there was no signal when RNasin was less than 0.2 U·μL⁻¹, and it was enhanced with the increased RNasin. The concentration of dNTPs was above 0.4 mmol·L⁻¹, the signal was appeared; the concentration of SuperScript II ranged from 0.1U·μL⁻¹ to 1.3

U µL$^{-1}$ and the signal was enhanced with the increase concentration of SuperScript II; the concentration of primers above 0.9 µmol·L$^{-1}$ were effective, less than 0.8 µmol·L$^{-1}$ could not synthesized sufficient quantities of cDNA and above 1.2 µmol·L$^{-1}$ could produce non-specific product.

### 4.5 The effect of other factors

The result showed that positive signals were appeared on the slide only when the annealing temperature at 56°C, which indicated that the suitable temperature was 56°C. Amplification with 10-20 cycles, the signals were not appeared, 25 cycles appeared weaker blue signal, 30-35 cycles showed stronger signals, which demonstrated that fewer cycles led to lower

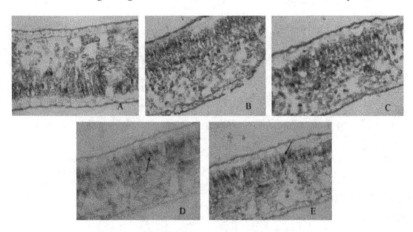

Fig. 2. The effect of cycle number on *In situ* RT-PCR
A: 10 cycles; B: 15 cycles; C: 20 cycles; D: 25 cycles; E: 30 cycles

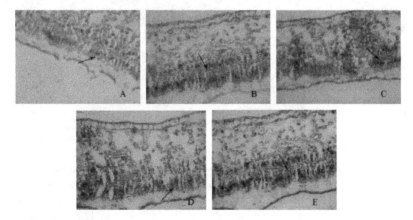

Fig. 3. The effect of the different *Taq* DNA polymerase concentration on the detection of *In situ* RT-PCR
A: 2 U ·100 µL$^{-1}$; B: 4 U ·100 µL$^{-1}$; C: 6 U ·100 µL$^{-1}$; D: 8 U ·100 µL$^{-1}$; E: 10 U ·100 µL$^{-1}$

synthesis (Figure. 2). The concentration of *Taq* DNA polymerase with 2 U ·100μL⁻¹-10 U ·100μL⁻¹ could satisfy amplification and showed stronger signals, which indicated that the suitable concentration of *Taq* DNA polymerase was 2 U ·100μL⁻¹ (Figure. 3).

### 4.6 PRINS-Rhodamine staining

Applied PRINS-Rhodamine staining detected ACLSV showed that the infected leaves of pear tissues were presented red fluorescence positive signals (Fig. 4, A~D, arrows showing the locations), which were consistent with the results of *In situ* RT-PCR detection (Niu et al., 2007). Healthy leaves and infected leaves without SuperScript II RT, fluorescent antibody and *Taq* DNA polymerase, did not present red fluorescence signals (Fig. 5, E~H).

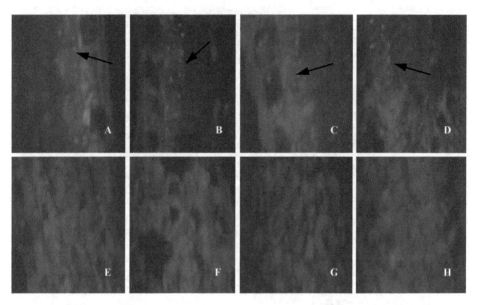

Fig. 4. PRINS-Rhodamine staining results of ACLSV in pear tissues
A-D: Labeled results of virus infected pear leaves from the same positions of different trees;
E: Labeled results of healthy pear leave (control); F-H: PRINS-Rhodamine staining results of ACLSV in pear tissues (control: Left out of SuperScript II RT, fluorescence antibody, *Taq* enzyme).

### 4.7 PRINS-FITC staining

FITC fluorochrome was more sensitive to the temperature and pH, and the efficiency was lower than Rhodamine staining, and the results showed inconspicuous signals. Applied PRINS-FITC staining detected ACLSV showed that the infected leaves of pear tissues were presented green fluorescence positive signals (Fig. 5, A~D, arrows showing the locations), which were consistent with the results of *In situ* RT-PCR detection (Niu et al., 2007). Healthy leaves and infected leaves without SuperScript II RT, fluorescent antibody and *Taq* DNA polymerase, did not present red fluorescence signals (Fig. 2, E~H).

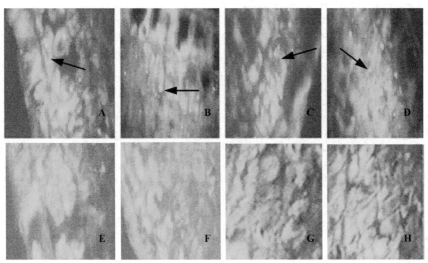

Fig. 5. PRINS-FITC staining results of ACLSV in pear tissues
A-D: Labeled results of virus infected pear leaves from the same positions of different trees;
E: Labeled results of healthy pear leave (control); F-H: PRINS- Rhodamine staining results of
ACLSV in pear tissues (control: Left out of SuperScript II RT, fluorescence antibody, *Taq*
enzyme).

## 5. Discussion

The study is based on virus RNA as a template to reverse transcription cDNA and *in situ*
amplification. Before amplification, the slides treatment with DNA exonuclease without
RNA enzyme overnight digest the original genomic DNA in tissues which can eliminating
DNA fragment decorated by polymerase which could form false-positive amplification
(Long et al., 1993). In our studies, the known virus-free material of pear tree used as the
negative control did not appear specificity of fluorescence signals. Negative control without
SuperScript II RT, fluorescence antibody, *Taq* enzyme showed the same result of virus-free
material. Signals did not display without RT steps indicated that the products were
amplified by cDNA, which excluded the possible of experimental reagents cross produced
fluorescent complex and attached to the tissue surface induced fluorescence signals. In our
studies, ACLSV of leave sections of Korla Pear were detected by *in situ* RT-PCR and PRINS,
the results showed that the positive materials were found obviously alcian blue and
fluorescence signals in mesophyll cells, while the negative control tissue did not appear. It
was indicated that ACLSV mainly distributed in the palisade tissue of the mesophyll cells,
and the same results as *in situ* RT-PCR detection (Niu et al., 2007). In addition, the results
showed that the thickness of section had a great influence on detection. Thin slices can easy
to cause the tissues were not complete, and the cell of thick slices were multiple and
overlapping, which unfavorable for observing, and seriously affect the detection results. So,
in order to obtain desire results of detection, the 4-6 μm of sections were used.

Because of the *in situ* amplified cDNA in tissues, we must consider the number of primers to
use. A single primer would not allow a strong enough signal for fluorescent detection.

However, too many primers would likely lead to primer-dimers or non-specific hybridization. In PRINS reaction system, primer extensions strictly followed the principle of complementary base pairing, and ensure the specificity labeling. Synthesis of labeled DNA will remain in the amplified position and not diffusion. In this study, we used five specific primers for PRINS, and achieved clearly fluorescence signals.

Terkelsen et al., (1993) using repeated primed *in situ* labeling (repeated-PRINS). This change of strategy results in a localized accumulation of sequence-specific labeled DNA, resulting in up to a 15-fold amplification of the signal as compared to the standard PRINS method. Ni et al., (1998) results showed that the repeated-PRINS technology could to enhance the signal; however, repeated heat denaturation and extension process for long time which induced the cell loss normal forms. In our study, we pretreatment species with appropriate concentration of protease K, and the optimal time of proteinase K digestion was necessary. The tissues slices were treated with proteinase K for 10, 20, 30, and 45 min. The best results were achieved after 20 min of the proteinase K digestion. The morphology of the tissue was well retained, and interpretation of results was unambiguous. The signal was recognized as fluorescence-signals the site of the label. The 10 min durations turned out to be too short and led to lack of signal. The extension of the reaction time up to 45 min produced morphological distortions to the point that interpretation of results became impossible. In addition, our research showed that increasing the ratio of dTTP and labeled-dUTP could improve the signal intensity. In general, the ratio of dTTP and labeled-dUTP was 1: 1 could generate enough strong signals. We increased the dTTP and labeled-dUTP concentration ratio to 1: 3 generated strong signals.

In this study, two fluorescence labeling were used, FITC and Rhodamine, respectively. Fluorescent-FITC was used *in situ* labeling showed sensitive on PH and easy to decay. In the conditions of susceptible pH or strong UV irradiation, the fluorescence excitation rapid decay. In addition, increase the times of washing, the tissues were more easily damaged and higher backgrounds were obtained. Therefore, on the basis of complete elution, appropriate to reduce washing processing steps were necessary.

Primed *in situ* labeling (PRINS) of nucleic acids was developed as an alternative to traditionally used fluorescence *in situ* hybridization (FISH). PRINS is based on sequence-specific annealing of unlabelled oligonucleotide primer under stringent conditions to the DNA of denaturated. Compared to FISH, PRINS is faster and does not require preparation of labeled probes, the process costs much less in terms of reagents (Velagelati et al., 1998; Tharapel & Kadandale, 2002; Pellestor et al., 2002), and hybridization signal is stronger, more specific and easy to control. In addition, we believe that this modified PRINS technique can have very meaningful applications in molecular cytogenetics. It can be used for the visualization and mapping of genetic loci on chromosomes, and for detection of the presence or absence of small DNA segments involved in genetic diseases. PRINS will have a more extensive application prospects in plant virus detection.

## 6. Conclusions

ACLSV of leave sections of Korla Pear were detected by *in situ* RT-PCR and PRINS, and the positive materials were found obviously alcian blue and fluorescence signals in mesophyll cells. The results showed that *in situ* RT-PCR and PRINS, which had two staining methods of PRINS-FITC and PRINS-Rhodamine, could get good detection results in which the parts have viruses showed alcian blue, green and red fluorescence light, respectively. Therefore, primed *in situ* labeling technique can be perfectly used for virus *in situ* detection of fruit trees, and it is also a rapid, simple and reliable *in situ* detection method.

## 7. Acknowledgements

This stsudy was supported by National Natural Science Foundation of China (30360066), the National Key Technologies R&D Program of China (2003BA546C), the Foundation Science and Technology Commission Xinjiang Production and Construction Crops, China (NKB02SDXNK01 SW) and Natural Science and Technology Innovation of Shihezi University, China (ZRKX200707).

## 8. References

Adams, M J.; Antoniw, J F.; Bar-Joseph, M.; Brunt, A A.; Candresse, T.; Foster, G D.; Martelli, G P.; Milne, R G. & Fauquet, C M. (2004). The new plant virus family Flexiviridae and assessment of molecular criteria for species demarcation. *Archives of Virology*, Vol.149, No.5, (May 2004), pp. 1045-1060, ISSN 0304-8608

Bagasra, O.; Hauptman, S. P.; Lischner, H. W.; Sachs, M, & Pomerantz, R. J. (1992). Detection of human immunodeficiency virus type 1 provirus in mononuclear cells by in situ polymerase chain reaction. *The New England Journal of Medicine*, Vol.326, No.21, (May 1992), pp. 1385-1391, ISSN 0028-4793

Chen, R. H. & Fuggle, S. V. (1993). In situ cDNA polymerase chain reaction. A novel technique for detecting mRNA expression. *American Journal of Pathology*, Vol.143, No.6, (December 1993), pp. 1527-1533, ISSN 0002-9440

Cieślińska, M.; Malinowski, T. & Zawadzka, B J. (1995). Studies on several strains of Apple chlorotic leaf spot virus (ACLSV) isolated from different fruit tree species. *Acta Horticulturae*, Vol.386, No.89, (July 1995), pp. 63-71, ISSN 0567-7572

Cohen, N. S. (1996). Intracellular localization of the mRNAs of argininosuccinate synthetase and argininosuccinate lyase around liver mitochondria, visualized by high-resolution in situ reverse transcription-polymerase chain reaction. *Journal of Cellular Biocemistry*, Vol.61, No.1, (April 1996), pp. 81-96, ISSN 1097-4644

Desvignes, J C. & Boyé, R. (1989). Different diseases caused by the Chlorotic leaf spot virus on the fruit trees. *Acta Horticulturae*, Vol.235, No.49, (April 1989), pp. 31-38, ISSN 0567-7572

Dunez, J.; Marenaud, G.; Delbos, R. P. & Lansac, M. (1972). Variability of symptoms induced by the *Apple chlorotic leaf spot* (CLSV). A type of CLSV probably responsible for bark split disease of prune trees, *Plant Disease Reporter*, Vol.56, No.4, (April 1995), pp. 293-295, ISSN 0032-0811

Greer, C. E.; Peterson, S. L.; Kiviat, N. B. & Manos M. M. (1991). PCR amplification from paraffin- embedded tissues. Effects of fixative and fixation time. *American Journal of Clinical Pathology*, Vol.95, No.2, (February 1991), pp. 117-124, ISSN 0002-9173

Gressens, P. & Martin, J. R. (1994). In situ polymerase chain reaction: Localization of HSV-2 DNA sequences in infections of the nervous system. *Journal of Virological Methods*, Vol.46, No.1, (January 1994), pp. 61-83, ISSN 0166-0934

Dunez, J. & Delbos, R. 1988, Apple chlorotic leaf spot virus, In: *European handbook of plant diseases*, I.M. Smith, (Ed.), 5-7, ISBN 0-632-01222-0, Britain

Haase, A. T.; Retzl, E. F. & Staskus, K.A. (1990). Amplification and detection of lentiviral DNA inside Cells. *Proceeding of the National Academy of Sciences of the United States of America*, Vol.87, No.13, (February 1990), pp. 4971-4975, ISSN 0027-8424

Harrer, T.; Schwinger, E. & Mennicke, K. (2001). A new technique for cyclic in situ amplification and a case report about amplification of a single copy gene sequence in human metaphase chromosomes through PCR-PR INS. *Human Mutation*, Vol. 17, No.2, (February 2001), pp. 131-140, ISSN 1059-7794

Höfler, H.; Pütz, B.; Mueller, J. D.; Neubert, W.; Sutter, G. & Gais, P. (1995). In situ amplification of measles virus RNA by the self-sustained sequence replication reaction. *Laboratory Investigation*, Vol.73, No.4, (October 1995), pp. 577-585, ISSN 0023-6837

Jelkmann, W. & Kunze, L. (1995). Plum pseudopox in German prune after infection with an isolate of Apple chlorotic leafspot virus causing plum line pattern. *Acta Horticulturae*, Vol.386, No.89, (July 1995), pp. 122-125, ISSN 0567-7572

Johansen, B. (1997). In situ PCR on plant material with sub-cellular resolution. *Annals of Botany*, Vol.80, No.5, (November 1997), pp. 697-700, ISSN 3035-7364

Kaczmarek, A.; Naganowska, B. & Wolko, B. (2007). PRINS and C-PRINS: Promising tools for the physical mapping of the lupin genome. *Cellular & Molecular Biology Letters*, Vol.12, No.1, (March 2007), pp. 16-24, ISSN 1425-8153

Koch, J.; Hindkjær, J.; Kølvraa, S. & Bolund, L. (1995). Construction of a panel of chromosome-specific oligonucleotide probes (PRINS-primers) useful for the identification of individual human chromosomes in situ. *Cytogenetics and Cell Genetics*, Vol.71, No.2, (June 1995), pp. 142–147, ISSN 0301-0171

Long, A. A.; Komminoth, P.; Lee, E. & Wolfe, H. J. (1993). Comparison of indirect and direct in-situ polymerase chain reaction in cell preparations and tissue sections. *Histochemistry*, Vol.99, No.2, (February 1993), pp. 151-162, ISSN 0948-6143

Lister, R M. (1970). Apple chlorotic leaf spot virus. In. *CMI/AAB Descriptions of Plant Viruses*, A.J. Gobbs (Ed.), 30, Commonwealth Mycological Institute, Kew, Surrey, England

Martelli, G. P.; Candresse, T. & Namba, S. (1994). *Trichovirus*, a new genus of plant viruses. *Archives of Virology*, Vol.134, No.3-4, (September 1994), pp. 451-455, ISSN 0304-8608

Németh, M. V. (1986). *Virus, Mycoplasma and Rickettsia diseases of fruit trees*, M. V. Németh, (Ed.), Akadémiai KiadóBudapest

Ni, B.; Li, H.; Zou, Y. H. & Wu, S. L. (1998). Use of the primed in situ labeling technique for a rapid detection of chromosomes X, 18. *Chinese Journal of Medical Genetics*, Vol.15, No.6, (December 1998), pp. 373-375, ISSN 1003-9406

Niu, J X.; Zhou, M. S.; Ma, B. G.; Zhao, Y. & Liu, H. (2007). Detection of Apple chlorotic leaf sopt virus in pear by in situ RT-PCR. *Acta Horticulturae Sinica*, Vol.34, No.1, (February 2007), pp. 53-58, ISSN 0513-353X

Nuovo, G. J.; MacConnell, P.; Forde, A. & Delvenne P. (1991). Detection of human papillomavirus DNA in formalin-fixed tissues by in situ hybridization after amplification by polymerase chain reaction. *American Journal of Pathology*, Vol.139, No.4, (October 1991), pp. 847-854, ISSN 0002-9440

Pellestor, F.; Imbert, I. & Andréo, B. (2002a). Rapid chromosome detection by PRINS in human sperm. *American journal of Medical Genetics*, Vol.107, No.2, (January 2002), pp. 109-114, ISSN 1552-4833

Pesquet, E., Barbier, O., Ranocha, P., Jauneau, A. & Goffner, D. (2004). Multiple gene detection by in situ RT-PCR in isolated plant cells and tissues. *The Plant Journal*, Vol.39, No.6, (September 2004), pp. 947–959, ISSN 1365-313X

Sato, K.; Yoshikawa, N. & Takahashi, T. (1993). Complete nucleotide sequence of the genome of an apple isolate of apple chlorotic leaf spot virus. *Journal of General Virology*, Vol.74, No.9, (September 1993), pp. 1927-1931, ISSN 0022-1317

Staskus, K. A.; Couch, L.; Bitterman, P.; Retzel, E. F.; Zupancic, M. & Haase, A. T. (1991). In situ amplification of visna virus DNA in tissue sections reveals a reservoir of latently infected cells. *Microbial Pathogenesis*, Vol.11, No.1, (July 1991), pp. 67-76, ISSN 0882-4010

Tharapel, S. A. & Kadandale, J. S. (2002). Primed in situ labeling (PRINS) for evaluation of gene deletions in cancer. *American journal of Medical Genetics*, Vol.107, No.2, (January 2002), pp. 123-126, ISSN 1552-4833

Tharapel, A. T. & Wachtel, S. S. (2006a). PRINS for mapping single-copy genes, In: *Gene Mapping, Discovery, and Expression*, M. Bina, (Ed.), 59-67, ISBN 1-58829-575-3, United States of America

Tharapel, A. T. & Wachtel, S. S. (2006b). PRINS for the detection of gene deletions in cancer, In: *PRINS and In Situ PCR Protocols*, F. Pellestor, (Ed.), 105-113, ISBN 1-58829-549-4, United States of America

Terkelsen, C.; Koch, J.; Kølvraa, S.; Hindkjr, J.; Pedersen, S. & Bolund, L. (1993). Repeated primer in situ labelling: Formation and labeling of specific DNA sequences in chromosomes and nuclei. *Cytogenetics and Cell Genetics*, Vol.63, No.4, (December 1993), pp. 235–237, ISSN 0301-0171

Velagelati, G. V.; Shulman, L. P.; Phillips, O. P.; Tharapel, S. A. & Tharapel, A. T. (1998). Primed in situ labeling for rapid prenatal diagnosis. *American journal of Obstetrics & Gynecology*, Vol. 178, No.6, (June 2001), pp. 1311-1320, ISSN ISSN: 0002-9378

Wachtel, S. S. & Tharapel, A. T. (2006). PRINS for the detection of unique sequences, In: *PRINS and In Situ PCR Protocols*, F. Pellestor, (Ed.), 33-40, ISBN 1-58829-549-4, United States of America

Woo, H. H., Brigham, L. A. & Hawes, M. C. (1995). In-cell RT-PCR in a single, detached plant cell. *Plant Molecular Biology Reporter*, Vol.13, No.4, (December 1995), pp. 355–362, ISSN 0734-9640

Xu, J.; Wang, A. & Chen, J M. (2002). New developments in primed in situ labeling (PRINS) methods and applications. *Journal of Clinical Medicine in Practice*, Vol.6, No.3, (February 2002), pp. 169-175, ISSN 1672-2353

Yan, J.; Bronsard, M. & Drouin, R. (2001). Creating a new color by omission of 3' end blocking step for simultaneous detection of different chromosomes in multi-PRINS technique. *Chromosoma*, Vol.109, No.8, (February 2001), pp. 565-570, ISSN 0009-5915

Yanase, H. (1974). Studies on apple latent viruses in Japan, *Bulletin of the Fruit Tree Research Station, Japan* Series C1, pp. 47-109 ISSN 0916-5851 0385-2326

Yoshikawa, N. & Takahashi, T. (1988). Properties of RNAs and proteins of apple stem grooving and apple chlorotic leaf spot viruses. *Journal of General Virology*, Vol.69, No.1, (January 1988), pp. 241-245, ISSN 0022-1317

# 4

# Analysis of Genomic Instability and Tumor-Specific Genetic Alterations by Arbitrarily Primed PCR

Nikola Tanic[1], Jasna Bankovic[1] and Nasta Tanic[2]

*[1]Institute for Biological Research "Sinisa Stankovic", University of Belgrade, Belgrade*
*[2]Institute of nuclear Sciences "Vinca", Belgrade*
*Republic of Serbia*

## 1. Introduction

It is now widely accepted that cancer development is a multistage process that results from an accumulation of mutations (Lengauer et al., 1998). Since spontaneous mutation rates in human cells are considerably lower then the large number of mutations observed in cancer cells, cancer cells must be a manifestation of the mutator phenotype. The mutator phenotype, also referred to as genomic instability, designates the increased mutation rate that occurs in neoplastic cells (Loeb, 1991). The induction of the genomic instability phenotype is emerging to be a crucial early event in carcinogenesis that enables an initiated cell to evolve into a cancer cell by achieving a greater proliferative capacity and genetic plasticity, which can overcome host immunological resistance, localized toxic environments and a suboptimal supply of micronutrients (Loeb, 1991; Cahill et al., 1999; Fenech 2002). Two distinct forms of genomic instability have been identified, microsatellite instability (MIN) and chromosomal instability (CIN). They probably encompass most characterized malignances (Lengauer et al., 1998; Breivik & Gaudernack, 1999). Genomic instability is present in all stages of cancer, from precancerous lesions to advanced cancers (Negrini et al., 2010; Markovic et al., 2008)

Measurements of instability have been performed by a variety of techniques, including flow cytometry, comparative genomic hybridization (CGH), allelotyping, and analysis of gene amplification rates (Vogelstein et al., 1989; Kallioniemi et al., 1994; Jass et al., 1994). These approaches, although informative, are generally cumbersome and somewhat impractical for widespread clinical use. Unlike these techniques, DNA fingerprinting methods, RAPD (Random Amplified Polymorphic DNA) and AP-PCR (Arbitrarily Primed Polymerase Chain Reaction) are rapid and simple procedures that examine the whole genome and detect the propensity of a tumor to undergo genomic rearrangements (Peinado et al., 1992; Perucho et al., 1996).

AP-PCR is a PCR-based DNA fingerprinting method that utilizes arbitrarily chosen primers to co-amplify multiple and independent sequences under low stringency conditions during the first cycles. It was first described by Welsh and McClelland (1990), who designed it to amplify multiple DNA fragments from anonymous regions of the genome. Initial cycles of the reaction are performed under low stringency conditions which are achieved with low temperatures during the annealing step of PCR and/or high magnesium concentration in the reaction. Under these conditions the arbitrary primer anneals to the best matches in the template. The priming events during the initial low stringency cycles are arbitrary since they depend on the nucleotide sequence of the PCR primer, which is arbitrarily chosen. Competition between these annealing events results in reproducible and quantitative amplification of many discrete bands. Further amplification of these sequences (discrete bands) under high stringency conditions produces a complex fingerprint which can be visualized by gel electrophoresis. The obtained band pattern is characteristic and representative of the genome used as template.

The large number of bands amplified with a single arbitrary primer generates a complex fingerprint that can be used to detect differences in the arbitrary amplified DNA sequences from two different but closely related genomes, like DNA from normal and cancer cells. Such differences correspond to somatic genetic alterations. In addition, AP-PCR method permits direct cloning and identification of altered variant bands i.e. altered DNA sequences. Therefore, this unbiased methodology allows for molecular karyotyping of somatically acquired genomic abnormalities, comparing related genomes, whereby one is a derivative of the other emerging via undefined and abnormal genomic events. Indeed, AP-PCR has been successfully used as a molecular alternative for cancer cytogenetics since it has proved to be capable of detecting chromosomal gains and losses as well as point mutations associated with carcinogenesis (Perucho et al., 1996; Chariyalertsak et al., 2005). This is based on the following favorable properties of the method: (i) the amplified bands usually originate from single copy sequences rather then from repetitive elements; (ii) there is no apparent bias for the chromosomal origins of the amplified bands, and therefore, fingerprints representative of the full chromosomal complement can be obtained by using a few arbitrary primers; (iii) the amplification is semi-quantitative, that is, the intensities of the amplified bands are almost proportional to the concentration of the corresponding template sequences.

Taking into account the potential and advantages of AP-PCR method, it seems as a reasonable approach to use this method to detect and quantify the level of genomic instability in various cancer samples. Therefore, we applied AP-PCR to measure genomic instability in samples of patients with Non Small Cell Lung Carcinoma (NSCLC) of various stages and grades, samples of patients with Malignant Gliomas of various grades (Anaplastic Astrocytomas and Glioblastomas) and samples of patients with Head and Neck Squamous Cell Carcinoma (HNSCC) and their premalignant lesions leukoplakias. Moreover, we aimed to identify some of these genomic alterations associated with the process of carcinogenesis in these types of tumors.

Here we describe the procedure for analyzing the level of genomic instability and identifying specific genetic alterations that occur during the tumorigenic process by Arbitrarily Primed PCR. This procedure involves the following steps: (i) comparative AP-PCR analysis of matching normal and tumor tissue and determination of the frequency of DNA alterations, a measurement of genomic instability; (ii) correlation between the level of genomic instability and histological grade and stage of each tumor; (iii) isolation and identification of altered amplified bands. Obtained results are presented and discussed in terms of the evolution of these types of tumors.

## 2. Materials and methods

### 2.1 Tissue samples and DNA extraction

Paired tumor and normal tissue samples (adjacent normal lung tissue and blood for malignant gliomas, HNSCC and leukoplakias) were analyzed. Specifically, 30 malignant glioma patients who underwent surgical resection at Clinic for Neurosurgery, Clinical Center of Serbia, 30 NSCLC patients who underwent surgery at the Institute for Lung Diseases and Tuberculosis, Clinical Centre of Serbia, 32 leukoplakia patients and 30 HNSCC patients who underwent surgery at the Clinic of Maxillofacial Surgery, School of Dentistry, University of Belgrade. Freshly excised tissue samples were partitioned for histopathology and DNA analyses. The specimens for DNA analyses were frozen in liquid nitrogen until DNA extraction. The samples were collected and used after obtaining informed consents and approval from the Ethics Committee, in accordance with the ethical standards laid down in the 1964 Declaration of Helsinki.

DNA was extracted using the phenol/chloroform/isoamyl alcohol method (Sambrook et al., 1989). The quality of the DNA was verified by electrophoresis on 0.8% agarose gel. The DNA concentration was assessed spectrophotometrically.

### 2.2 AP-PCR DNA fingerprinting

Genomic instability was determined by comparing the AP-PCR profiles of paired tumor and normal DNA samples of the same patient. Altogether, twenty primers were tested for the ability to generate informative fingerprints that distinguish tumor from normal tissue. Optimization of AP-PCR reactions was done for each primer according to Cobb (1997) and included the search for conditions that would yield profiles of moderate complexity in order to simplify the analysis (McClelland & Welsh, 1994). Normally, optimization of AP-PCR DNA fingerprinting would require each variable to be tested independently. An experiment investigating the effects and interactions of four critical reaction components (dNTPs, $MgCl_2$, primer and DNA), each at three concentrations, would require 81 (i.e., $3^4$) separate reactions. However, using modified Taguchi method (Taguchi & Wu, 1980, as cited in Cobb, 1997) only nine reactions are required to perform the same optimization. Here an estimate of the effect of individual components is achieved by looking at the effects that component interactions have on the fingerprint. These interactions are determined by arranging those components that are likely to affect the reactions into an orthogonal array. The product yield for each reaction is used to estimate the effects of individual components on amplification. We varied the PCR components in the following final concentrations: dNTPs (0.2 mM, 0.4 mM, 0.6 mM), $MgCl_2$ (1.5 mM, 2.5 mM, 3.5 mM), primer (1.5 $\mu$M, 3.0 $\mu$M, 5.0 $\mu$M) and DNA (50 ng, 100 ng, 150 ng). DNA concentration did not affect AP-PCR fingerprints and it was used to validate the method. Namely, after optimal reaction conditions were established, each experiment included the analysis of two template concentrations (25 ng and 50 ng in a final volume of 25 $\mu$L) for each individual in order to exclude artifacts arising from impurities in the DNA preparations.

Twelve out of twenty primers produced informative profiles differentiating normal from tumor tissue. Primer sequences, AP-PCR conditions and reaction mixtures are given in Table 1.

| Primer | Primer sequence | AP-PCR low-stringency conditions | AP-PCR high-stringency conditions | AP-PCR reaction mixture |
|---|---|---|---|---|
| CCNA1 | 5'-AAG AGG ACC AGG AGA ATA TCA-3' | 95°C 30" 45°C 2' 72°C 1' | 95°C 30" 60°C 1' 72°C 1' | 0,2mM each dNTP; 3,5mM MgCl₂; 5μM primer; 1U Taq DNA |
| LRP-A | 5'-GCT TCC GAG GTC TCA AAG C-3' | 95°C 30" 40°C 2' 72°C 1' | 95°C 30" 58°C 1' 72°C 1' | 0,2mM each dNTP; 3,5mM MgCl₂; 5μM primer; 1U Taq DNA |
| MDR-A | 5'-GTT CAA ACT TCT GCT CCT GA-3' | 95°C 30" 40°C 2' 72°C 1' | 95°C 30" 58°C 1' 72°C 1' | 0,4mM each dNTP; 2,5mM MgCl₂; 5μM primer; 1U Taq DNA |
| E8S p53 | 5'-TAA ATG GGA CAG GTA GGA CC-3' | 95°C 30" 40°C 2' 72°C 1' | 95°C 30" 58°C 1' 72°C 1' | 0,4mM each dNTP; 2,5mM MgCl₂; 5μM primer; 1U Taq DNA |
| GAPDH-S | 5'- CGG AGT CAA CGG ATT TGG TCG TAT-3' | 95°C 30"; 50°C 2'; 72°C 1' | 95°C 30"; 70°C 1'; 72°C 1' | 0,4mM each dNTP; 2,5mM MgCl₂; 5μM primer; 1 U Taq DNA |
| GAPDH-A | 5'-AGC CTT CTC CAT GGTGGT GAA GAC-3' | 95°C 30"; 50°C 2'; 72°C 1' | 95°C 30"; 72°C 1'; 72°C 1' | 0,2 mM each dNTP; 2,5 mM MgCl₂; 3 μM primer; 1 U Taq DNA |
| E5A p53 | 5'-CAG CCC TGT CGT CTC TCC AG-3' | 95°C 30"; 40°C 2'; 72°C 1' | 95°C 30"; 55°C 1'; 72°C 1' | 0,6 mM each dNTP; 3,5 mM MgCl₂; 3 μM primer; 1 U Taq DNA |
| p53 A | 5'-TTG GGC AGT GCT CGC TTA GT-3' | 95°C 30"; 40°C 2'; 72°C 1' | 95°C 30"; 60°C 1'; 72°C 1' | 0,2 mM each dNTP; 3,5 mM MgCl₂; 5 μM primer; 1 U Taq DNA |
| H61-5' | 5'-AGG TGG TCA TTG ATG GGG AG-3' | 94°C 1'; 45°C 2", 72°C 2' | 94°C 1'; 62°C 1'; 72°C 2' | 0.4 mM each dNTPs, 2.5 mM MgCl₂ 5 μM primer; 1 U Taq DNA |

Table 1. Primer sequences, AP-PCR conditions and reaction mixtures

The reactions consisted of an initial denaturation step (95°C for 5 min), 4 cycles at low-stringency conditions (specific for each primer), 35 cycles at high-stringency conditions (specific for each primer), and a final extension (72°C for 7 min) in a GeneAmp® PCR System 9700 (Applied Biosystems, Foster City, California, USA).

The AP-PCR products were separated on 6 – 8% non-denaturing polyacrylamide (PAA) gels and visualized by silver staining. Silver- staining procedure creates permanent record of the electrophoresis results and includes several steps: fixing, silver impregnation, development and stopping the reaction. In the fixing step, the gel is treated with 1% nitric acid solution to render the macromolecules in the gel insoluble and prevent diffusion during the subsequent staining steps. In the silver impregnation step, soluble silver ion (Ag+) derived from the silver nitrate, 12mM $AgNO_3$ solution, binds to nucleic acid bases fixed in gel. Generally, DNA bases promotes the reduction of silver ion to metallic silver (Ag0), which is insoluble and visible, allowing nucleic acid-containing bands to be seen. In order to prevent reduction of silver ion to metallic silver before the end of silver impregnation, this step is often performed in mildly acidic acid conditions. During the development step, formaldehyde reduces silver ions to metallic silver in process that only proceed at high pH, approximately 12. For that reason, sodium carbonate is included as one of the main component that render development solution alkaline. Stopping reaction step imply prevention of any further silver ion reduction by soaking the gels in the 10 % acetic acid solution. Finally, it should be emphasized that water washes are also included between some of the above mentioned steps in the silver staining procedure (detailed procedure is given in Table 2).

| Step | Solution | Time |
|---|---|---|
| Fixation Pretreatment | 10% Ethanol | 10 minutes |
| Fixation | 1% Nitric Acid Solution | 3 minutes |
| Water Washing | Destiled H2O | 2 x 1 minute |
| Silver Impregnation | 12 mM Silver Nitrate Solution | 30 minutes |
| Water Washing | Destiled H2O | 3 x 1 minutes |
| Developing - Reduction | 0.28 M Sodium Carbonat with 0.019 % Formaldehyde | Until desired images appear |
| Stopping Reduction | 10 % Acetic Acid | 5 – 10 minutes |

Table 2. In-house procedure for silver- staining of PAA gels.

Gel images were acquired with the Multi-Analyst/PC Software Image Analysis System (Bio Rad Gel Doc 1000). Digitized images were loaded into the specialized public software Image J (National Institute of Health, USA, www.rsb.info.nih.gov/ij) and analyzed by the image enhancement function 'adapthisteq'. This function performs contrast-limited adaptive histogram equalization on small regions of the image, called tiles. Contrast of each tile is enhanced so that the histogram of the output region approximately matches a specified histogram. After equalization, adapthisteq combines neighboring tiles using bilinear interpolation to eliminate artificially induced boundaries.

## 2.2.1 Reproducibility

The problem of reproducibility of AP-PCR has been a matter of concern for quite some time (Meunier and Grimont, 1993; McClelland and Welsh, 1994). In our study, reproducibility was verified by at least three independent reactions and a reaction with a two-fold higher template concentration. Occasional irreproducibility was found to be due to template quality, where additional round of purification solved the problem. Template carry-over was routinely monitored by systematic incorporation of "no-template reaction" in each set of experiments. Day to day variation was found only in respect of band intensities. This variability was in the range of less than 10% (± 5%) as estimated by integration of densitometric scans. Interlab variation was not assessed but we presume that it does not affect the interpretation of data from this report.

## 2.3 Isolation, cloning and DNA sequencing of variant bands obtained by AP-PCR

Selected variant DNA bands, bands with altered mobility, were further characterized. The PCR amplicons resolved on the silver stained gels were gently removed with a hypodermic 22-gauge needle pre-wetted with the PCR master mix solution. The needle was dipped in the PCR master mix for 2 min and then discarded. The PCR products were reamplified with the same primers used for AP-PCR reactions at high-stringency conditions specific for each particular primer. The reamplified material was administrated on 1.5% agarose gels, purified using DNA Extraction Kit (Fermentas Life Sciences, Lithuania) and cloned with GeneJetTM PCR Cloning Kit (Fermentas Life Sciences, Lithuania) according to manufacturers' instructions. Plasmids were purified using GeneJetTM Plasmid Miniprep Kit (Fermentas Life Sciences, Lithuania).

Cloning process consisted of setting up the blunting and ligation reactions. Blunting reaction allows the conversion of PCR products generated with non-proofreading Taq DNA polymerase to DNA fragments with blunt ends using thermostable DNA Blunting Enzyme provided with the kit. The reaction consists of 10 µL of 2x Reaction Buffer, 2 µL of non-purified PCR product, 5 µL of nuclease free water and 1 µL of DNA Blunting Enzyme in 18 µL reaction mixture. The resulting blunt-ended DNA can be ligated efficiently into a vector, pJET1.2/blunt, using the included DNA Ligation Kit Solutions: 1 µL of pJET1.2/blunt Cloning Vector (50ng/ µl) and 1 µL of T4 DNA Ligase (5u/µl). The vector contains a lethal restriction enzyme gene that is disrupted by ligation of a DNA insert into the cloning site. As a result, only bacterial cells with recombinant plasmids are able to form colonies. Recircularized pJET1.2/blunt vector molecules lacking an insert express a lethal restriction enzyme which kills the host E.coli cell after transformation. This positive selection drastically accelerates the process of colony screening and eliminates additional costs required for blue/white selection. The reactions can be used directly for bacterial transformation and in vitro packaging procedures without further purification. All common laboratory E.coli strains can be directly transformed with the ligation product.

Before the transformation procedure, the preparation of competent bacteria of *E. coli* GM2163 strain was performed using TransformAid™ Bacterial Transformation Kit (Fermentas Life Sciences, Lithuania) according to the manufacturer instruction.

The next step was to recover plasmid DNA from recombinant *E.coli* cultures using GeneJET™ Plasmid Miniprep Kit (Fermentas Life Sciences, Lithuania). A single colony was picked from a freshly streaked selective plate for inoculation of 5 mL of LB liquid medium (Fermentas Life Sciences, Lithuania) supplemented with the ampicillin. A bacterial culture is harvested and lysed. The lysate is then cleared by centrifugation and applied on the silica column to selectively bind DNA molecules at a high salt concentration. The adsorbed DNA is washed to remove contaminants, and the pure plasmid DNA is eluted in a small volume of elution buffer or water. The purified DNA is ready for immediate use in all molecular biology procedures such as automated sequencing. Before sequencing, the ligation of DNA fragment into the plasmid was verified using restriction enzymes HindIII and EcoRI (Sigma-Aldrich Chemie GmbH, Germany). The fragments obtained after restriction were analyzed on 1% agarose gels. The sequencing was performed only after the presence of the DNA fragment in plasmid was confirmed by comparing the molecular weight of recombinant plasmid with DNA ladder.

Sequences were determined on ABI Prism 3130 Genetic Analyzer automated sequencer (Applied Biosystems, Foster City, CA, USA) using BigDye Terminator v3.1 Cycle Sequencing Kit (Applied Biosystems, Foster City, CA, USA). Sequencing was performed in both directions on several clones for each selected DNA band. The obtained sequences were analyzed using BLAST software in the NCBI GenBank and EBI (Sanger Institute) database.

The sequencing procedure itself involved: 1) two independent cycle sequencing PCRs, each with one primer only (5′ and 3′), for the sequencing in both directions; 2) precipitation of the amplicons; 3) their denaturation and 4) automatic electrophoresis. Cycle sequencing PCRs were performed on the GeneAmp® PCR System 9700 (Applied Biosystems, Foster City, CA, USA) using BigDye Terminator v3.1 Cycle Sequencing Kit (Applied Biosystems, Foster City, CA, USA) with the final concentration of 100-300 ng of the plasmid DNA and 4pmol of the primer under the following conditions: initial denaturation at 96°C for 1 min, 25 cycles at 96°C for 10 s, 50°C for 5 s, 60°C for 4 min and at 4°C indefinitely. The obtained PCR products were precipitated and EDTA (25 mM final) and EtOH (70-75% final) added.The mixture was incubated for 15 min at RT and then centrifuged 30-45 min at 6000 rpm and +4°C. The supernatant was removed, a new quantity of 70% EtOH added, followed by centrifugation for 25 min at 5000 rpm and +4°C. Supernatant was removed again and the obtained pellet dried at 90°C. Then, 15 μl of Hi-DI™ Formamide (Applied Biosystems, Foster City, CA, USA) was added for the denaturation at 95°C. 10 μl of the amplicons dissolved in formamide were subjected to the automatic electrophoresis and sequence reading on ABI Prism 3130 Genetic Analyzer automated sequencer (Applied Biosystems, Foster City, CA, USA). The obtained sequences were analyzed using BLAST software in the NCBI GenBank and EBI (Sanger Institute) database.

## 3. Results and discussion

Genomic instability was determined by comparing the AP-PCR profiles of DNA isolated from paired normal and tumor tissues of patients with non small cell lung cancer (NSCLC), malignant glioma, head and neck squamous cell carcinoma (HNSCC) and leukoplakia (L). Twelve out of twenty tested primers produced informative amplification profiles differentiating normal from tumor tissue or normal from leukoplakia (Table 1). Specifically, five primers produced informative sequence alterations that distinguish NSCLC from normal tissue, a set of four primers produced informative fingerprints differentiating malignant gliomas from normal tissue and another set of four primers produced informative sequence alterations that distinguish HNSCC and leukoplakias from their normal counterparts. The AP-PCR products were separated on 6-8% nondenaturing polyacrylamide (PAA) gels and visualized by silver staining. Typical fingerprints are shown in Figure 1.

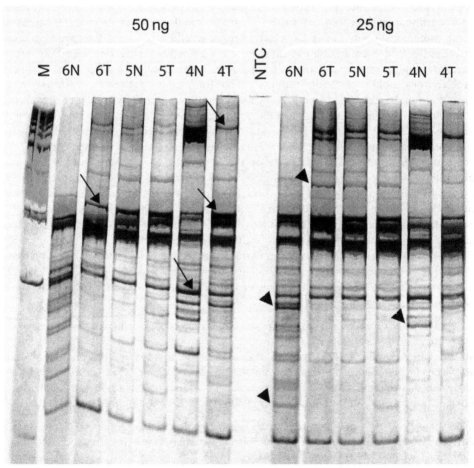

Fig. 1. AP-PCR fingerprint profiles of tumor (T) and normal (N) tissues from patients with NSCLC obtained with GAPDH AS primer. Reactions were performed in duplicate with 25 ng and 50 ng of DNA. Numbers 1–5 represent the patients; M–the DNA ladder; NTC–no template control. Arrows and arrowheads indicate examples of quantitative and qualitative changes, respectively.

This type of analysis differentiates individuals and, thus, displays the cardinal feature of the DNA profile analysis. Additionally, some bands are characteristic for the human genome, being common to all analyzed patients. Importantly, some electrophoretic bands were present in DNA profiles of tumor but not in normal tissue, and vice versa, indicating the mutational like events. The unbiased nature of AP-PCR profiling allows for the screening of anonymous regions of a genome without any prior knowledge of its structure (Welsh and McClelland, 1990; Williams et al., 1990) and provides information about two distinct types of DNA alterations: qualitative and quantitative. Qualitative differences, which represent microsatellite instability (MIN), are detected as mobility shifts in the banding pattern, i.e., the presence or absence of specific bands in tumor and control samples. Quantitative differences appear as altered band intensities and represent amplifications or deletions of existing chromosomal material as manifestations of chromosomal instability (CIN). Observed changes should be cautiously regarded as semi-quantitative and semi-qualitative due to the competitive nature of AP-PCR where sequence context may play unpredictable role. This situation may present a serious problem for simple to moderate patterns but not for complex patterns. Unfortunately, the former are preferred due to simplicity of interpretation. Since the profile is the result of a competition between many PCR products, the problem may appear with very simple profiles in analysis of similar but non-identical genomes. For this reason, it had been suggested to use profile pattern with more than 10 prominent PCR products of moderate complexity (McClelland and Welsh, 1994). We followed this reasoning and the necessary precautions for reproducibility and reliability of DNA profiling analysis in comparing DNA fingerprints of paired normal – tumor samples. We identified significant genomic instability in most cases as qualitative and quantitative electrophoretic changes. The qualitative alterations represented as a loss or a gain of a band are the result of mutations at the primer-template interaction sites leading to a mobility shift of a band. Quantitative changes were observed as bands of either decreased or increased intensity. Allelic losses, which may occur as a result of their linkage to suppressor genes, produce bands with decreased intensity. Gene amplification or chromosomal aneuploidy appears as bands with increased intensity.

For each type of DNA change, as well as for the total number of changes, the frequency of DNA alterations, a measurement of genomic instability, was calculated as the number of altered bands in the AP-PCR profile of tumor tissue divided by the total number of amplicons in the fingerprint of normal tissue from each patient. AP-PCR fingerprints were analyzed and qualitative and quantitative changes determined using image enhancement function 'adapthisteq' of the specialized public software Image J (Figure 2).

DNA alterations were detected in all analyzed samples with the frequency varying among different types of tumors (Table 3). The largest variation of the frequency of total DNA alterations was in NSCLC patients ranging from 8% to even 68%. The contribution of qualitative changes to overall genomic instability was significantly greater than the contribution of quantitative changes. This large range of instability raised the question of its distribution among samples of NSCLC patients. In other words we were interested to see if there was association between the level of genomic instability and any clinicopathological parameter.

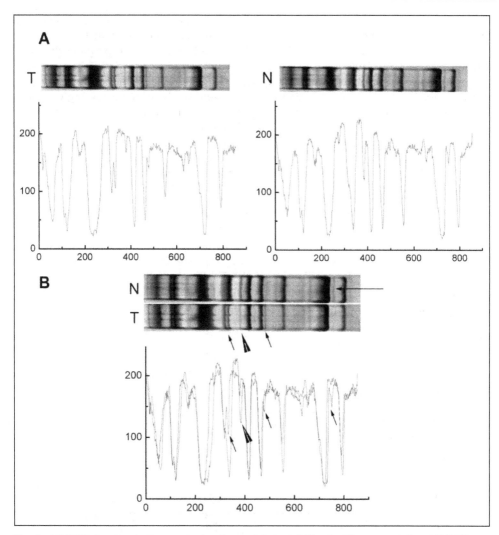

Fig. 2. AP-PCR fingerprinting analysis of genomic instability in glioma samples. AP-PCR profiles of tumor (T) and blood (N) tissues from the same patient obtained using MDRa primer, separated on 6% non-denaturing polyacrylamide (PAA) gel and corresponding contrast-limited adaptive histograms obtained using image enhancement function 'adapthisteq' of the specialized public software Image J (**A**). Arrows and arrowheads indicate examples of qualitative and quantitative electrophoretic changes respectively, clearly seen on the overlap of tumor and blood histograms (**B**).

The most noteworthy finding of this study was the association between the level of genomic instability and histological grades of NSCLC. Namely, we found the significant decrease of the total number of DNA alterations with increasing histological grade of the NSCLC. The same pattern was found for quantitative changes alone – the frequency of alterations decreased with the increase of the histological grade (Figure 3). These results support the idea that mutational alterations conferring genomic instability and the mutator phenotype occur early during tumor formation. The mutator phenotype hypothesis proposes that such phenotypes result from mutations in genes that maintain genomic stability in normal cells. Instability promotes mutations in other genes, oncogenes and tumor suppressor genes, providing the tumor cell with a selective growth advantage. These findings strongly support the increasingly popular explanation of neoplastic transformation in terms of Darwinian evolutionary mechanisms (Breivik, 2001; Breivik & Gaudernack, 1999; Cahill et al., 1999). Evolution through natural selection depends on two essential elements, the availability of

| type of DNA alteration | frequency of DNA alterations | | | |
|---|---|---|---|---|
| | NSCLC | glioma | leukoplakia | HNSCC |
| qualitative | 0.07 – 0.53 | 0.06 – 0.27 | 0.18 – 0.41 | 0.05 – 0.21 |
| quantitative | 0.01 – 0.16 | 0.05 – 0.27 | 0.07 – 0.16 | 0.07 – 0.21 |
| TOTAL | 0.08 – 0.68 | 0.14 – 0.49 | 0.30 – 0.48 | 0.12 – 0.31 |

Table 3. Measurement of genomic instability in various types of tumors.

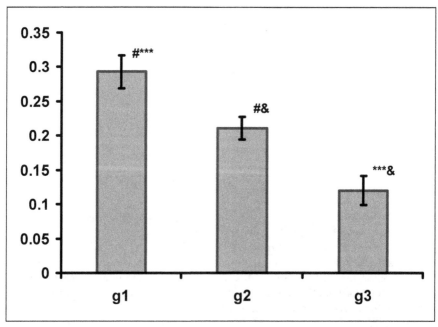

Fig. 3. The relationship between the total frequency of DNA alterations and the histological grades of the lung tumors. All values are presented as means ± SEM. # $p < 0.05$ when grade 1 was compared to grade 2; *** $p < 0.005$ when grade 1 was compared to grade 3; & $p < 0.05$ when grade 2 was compared to grade 3.

genetic variation and selection pressure (Dawkins, 1989.). In general evolutionary terms, it could be said that genomic instability accelerates the somatic evolutionary process by promoting genetic variation in an organism. Extensive genomic instability is thus expected in early phases of cancer progression (histological grade 1 in this study). At the same time, an increased mutation rate is expected to cause mutations that are deleterious or lethal at higher frequencies rather than mutations that have favorable effects on cellular proliferation. Consequently, elevated mutation rates must generally be regarded as disadvantageous to cellular growth (Tomlinson et al.,1996). Theoretical arguments suggest that the accumulation of large numbers of mutations can exceed the error threshold for cell replication and viability (Eigen, 1993). Only cells carrying reasonable number of mutations with favorable effects on cell growth would survive. Therefore, it seems probable that the expression of the mutator phenotype could be decreased and lost in the late phases of tumor progression. As a result, tumors may no longer exhibit a mutator phenotype but will nevertheless reveal its history, i.e. random mutations, throughout their genome (Loeb, 2001). In other words, the result showing the lower degree of genomic instability in advanced NSCLCs (grades 2 and 3) is not unexpected in the light of these arguments and could be considered as a marker of poor prognosis.

Following the study of genomic instability in NSCLC tissue samples, we made an attempt to identify some of detected DNA changes in order to identify genes that alter during NSCLC promotion and progression (Bankovic et al., 2010). Selected DNA bands with altered mobility were further characterized. Twenty one unique bands present only in tumor but not in normal tissue were retrieved from the gels and cloned. Variant bands that appeared in more than one sample (new bands with the same mobility), were chosen in order to identify DNA alterations common to as many NSCLC patients as possible. Bands (amplicons) with the same electrophoretic mobility were isolated and characterized from at least two patients in order to confirm that they represent the same DNA sequence. Three clones of each band were sequenced. Obtained sequences were submitted to homology or identity search in NCBI GenBank and EBI (Sanger Institute) databases. Following genes were identified: tetraspanin 14 (TSPAN14), cadherin 12 (CDH12), retinol dehydrogenase 10 (RDH10), cytochrome P450, family 4, subfamily Z, polypeptide 1 (CYP4Z1), killer cell immunoglobulin-like receptor (KIR), E2F transcription factor 4 (E2F4), phosphatase and actin regulator 3 (PHACTR3), PHD finger protein 20 (PHF20), PRAME (preferentially expressed antigen in melanoma) family member and solute carrier family 2 (facilitated glucose transporter), member 13 (SLC2A13). Moreover, we were able to identify types of mutations in revealed genes according to sequence data and BLAST search results and to examine their presence in relation to NSCLC subtype, histological grade and stage of the tumor, lymph node invasion and patients' survival. Examining their relation to the patients' clinicopathological parameters and survival we concluded that TSPAN14, SLC2A13 and PHF20 could have a role in NSCLC promotion, CYP4Z1, KIR and RDH10 would possibly play a role in NSCLC progression, while E2F4, PHACTR3, CDH12 and PRAME family member probably play important role in NSCLC geneses. Patients with altered E2F4 and PHACTR3 lived significantly shorter.

Unlike NSCLC samples, all leukoplakias demonstrated extensive instability in a relatively small range (Table 3). The frequency of total DNA alterations ranged from 0.30 to 0.48 and clearly distinguished two groups of leukoplakias: a group of six leukoplakias had a frequency of DNA alterations of 0.3–0.34 and was denoted as leukoplakias with a moderate degree of instability while the other group of 26 leukoplakias had a frequency of DNA alterations of > 0.4 and was denoted as leukoplakias with a high degree of instability (Tanic et al., 2009). However, such high levels of genomic instability in leukoplakia samples were a surprise mainly because they are defined as white patches or plaques of oral mucosa that cannot be rubbed off and cannot be diagnosed clinically or pathologically as other specific diseases and have been considered premalignant lesions only since recently (Neville & Day, 2002; Hunter et al., 2005). It is impossible to state, with precision, the proportion of leukoplakias that undergo malignant transformation. For oral mucosa, in general, up to 20% of leukoplakias exhibit dysplasia. Dysplastic leukoplakias have a greater probability of developing into cancer, although leukoplakias without evidence of dysplastic changes may also progress to highly aggressive squamous cell carcinoma. Still, the majority of leukoplakias fail to undergo malignant transformation. The frequency of malignant alterations in oral leukoplakia varies from study to study and ranges from 8.9 to 17.5% (summarized in Neville & Day, 2002). These facts and our finding were the reasons to include samples of Head and Neck Squamous Cell Carcinoma patients, identify and quantify genomic instability in these samples and compare obtained results with those of leukoplakia samples.

Obtained frequency of DNA alterations in HNSCC samples was significantly lower than that of leukoplakia samples, as shown in Table 3. When comparing mean frequencies of DNA alterations the result is even more convincing. Namely, mean frequency of total DNA changes was 0.42 for leukoplakia samples vs. 0.28 for HNSCC samples. Interestingly, contribution of quantitative changes to the total instability in HNSCC samples is significantly higher (0.21) than the contribution of qualitative changes (0.16) which is quite opposite in leukoplakia samples. In other words, the level of genomic instability decreased during HNSCC promotion from premalignant lesions but more serious alterations, quantitative changes as manifestations of chromosomal instability, were selected. These results fit nicely into Darwinian evolutionary theory of neoplastic transformation. High instability is present at the very beginning of HNSCC genesis, providing genetic variability in the population of premalignant cells, which is absolutely necessary for the evolution by natural selection. During tumor progression the level of instability decreases due to selection of genotypes that are better adapted to the micro-environment in which natural selection took place. However, the question remains: why the majority of leukoplakias with such a huge instability fail to undergo malignant transformation? The answer may be in exceeding the error threshold for cell replication and viability (Eigen, 1993) with so many mutations. In other words, it seems that leukoplakias with a high degree of genomic instability have less chance to develop into HNSCC, whereas leukoplakias with a lower (moderate) degree of genomic instability have a better chance of transforming, probably because they carry a certain number of mutations that have favorable effects on cell growth (Tanic et al., 2009).

Following the same reasoning as in the case of NSCLC we attempted to identify some of detected DNA changes in leukoplakias, with the aim of identifying tumor-specific alterations (Peinado et al., 1992) that could lead to the development of potential diagnostic markers involved in the genesis of HNSCC. To that end, nine variant bands present in leukoplakias but not in normal tissue, were selected. Unexpectedly, two different amplicons, originating from distinct leukoplakias, were identified as altered part of the TIMP-3 gene (tissue inhibitors of metalloproteinases 3), two were identified as mutated DNMT 3A gene (DNA (cytosine-5)-methyltransferase 3 alpha) and two represented copies of the Ty1-copia-like retrotransposon.

Further investigations of the detected genes in both, leukoplakia and NSCLC samples, on larger sample size, with special emphases on tumor promoting genes, are underway. We expect more detailed profile of their involvement in NSCLC and HNSCC after extensive analyses of their mutational status and detailed analyses of their expression profile at RNA and protein level in a larger sample. We expect that some of them might prove to be a good prognostic biomarkers for NSCLC or HNSCC patients.

Finally, we analyzed malignant gliomas, tumors that originate from glia, the most common and deadly brain tumors. All patients had histologically confirmed diagnosis of anaplastic astrocytoma (AA) or glioblastoma multiforme (GBM) according to the new World Health Organisation (WHO) classification. Anaplastic astrocytomas (WHO grade III) and glioblastomas (WHO grade IV) are two major groups of malignant gliomas. Glioblastomas are further classified as primary and secondary. Distinction between them is based on different genetic pathways leading to their development (Ohgaki & Kleihues, 2007; Van Meir et al., 2010). Primary glioblastoma develop rapidly *de novo*, without clinical or histological evidence of a less malignant precursor lesion. Secondary glioblastoma develop slowly progressing from low-grade diffuse astrocytoma (WHO grade II) or anaplastic astrocytoma (WHO grade III).

Examination of the extent of genomic instability revealed that samples of patients with anaplastic astrocytoma had similar level of total, microsatellite and chromosomal genomic instability as patients with glioblastoma multiforme, with very high values in both histological subtypes (Table 4). It was unexpected and, at first sight, looked like these results contradicted the expectation and results obtained from NSCLC and HNSCC samples. However, all analyzed grade IV glioblastomas were classified as primary glioblastomas (because glioblastoma diagnosis was made at the first biopsy, without clinical or histopathologic evidence of a less malignant precursor lesion), which are considered to be *de novo* tumors and not the progressive form of grade III astrocytomas. Therefore, obtained results are still consistent with the evolutionary theory of neoplastic transformation and the decrease of the level of genomic instability could be expected in secondary glioblastomas. In other words, extensive genomic instability might be used as diagnostic character where pathology cannot provide unambiguous distinction between primary and secondary GBM. Similar results were obtained by Nishizaki et al. (2002) who demonstrated that there was no significant difference in FISH heterogeneity between malignant gliomas of WHO grades III and IV. We expect that further research involving secondary glioblastomas will confirm our hypothesis and will provide additional confirmation for the evolutionary theory of tumor progression. Moreover, we hope that cloning and sequencing of amplified DNA bands showing genetic alterations specific for glioma genome, will allow the detection of new genes implicated in glioma pathogenesis and progression.

| instability | Mean frequency | | |
|---|---|---|---|
| | Anaplastic Astrocytoma | vs. | Glioblastoma Multiforme |
| microsatellite | 0.15 | vs. | 0.16 |
| chromosomal | 0.19 | vs. | 0.16 |
| TOTAL | 0.34 | vs. | 0.33 |

Table 4. Mean frequency of DNA alterations in malignant glioma samples.

Finally, it is worth mentioning that measurements of genomic instability could be performed by another DNA fingerprinting technique, RAPD (Random Amplified Polymorphic DNA). Wang et al. (2002) measured genomic instability in various cancer types using RAPD and the instability they detected was in average higher than 40% for lung cancer tissues. In another study (Ong et al., 1998), DNAs from 20 lung cancer (18 non-small cell lung cancers and two small cell lung cancers) and their corresponding normal tissues were amplified individually by RAPD with seven different 10-base arbitrary primers. PCR products from RAPD were electrophoretically separated in agarose gels and banding profiles were visualized by ethidium bromide staining. The ability to detect genomic instability in 20 cancer tissues by each single primer ranged from 15 to 75%. DNA changes were detected by at least one primer in 19 (95%) cancer tissues. They concluded that these results seem to indicate that genomic rearrangement is associated with lung carcinogenesis and that RAPD analysis is useful for the detection of genomic instability in lung cancer tissues.

Misra A. et al. (2007) used RAPD to attempt to quantify the number of clonal mutations in primary human gliomas of astrocytic cell origin . They targeted genomic loci of a different nature and estimated that the number of overall alterations in tumor genome seemed to be greater than expected. They also observed a higher number of genetic changes in tumors of lower grade and suggested that it could be a consequence of an increased mutation rate in early tumorigenesis due to acquisition of a mutator phenotype. The increased extent of alterations occurring in tumors of a lower grade is consistent with our study. The results of Misra et al. showed the acquisition of a mutator phenotype early in tumorigenesis and support the mutator hypothesis proposed by Loeb (1991, 2001).

## 4. Conclusions

AP-PCR DNA fingerprinting is an efficient tool to quickly and easily screen a very large number of loci for possible DNA alterations in cancer cells. It has several advantages: first, minor amounts of template DNA are sufficient for analysis; second, it allows for the screening of anonymous regions of a genome without any prior knowledge of its structure; third, two types of DNA alterations could be detected in single reaction, chromosomal rearrangements and random mutations dispersed over the genome; and forth, possibility of reamplification, cloning and sequencing of variant bands enables the rapid identification of the genes probably linked to tumor progression. Here, we demonstrated the use of AP-PCR DNA fingerprinting in detection and quantification of genomic instability (microsatellite, chromosomal and total) in three types of tumors as well as in search for molecular biomarkers for cancer promotion and progression. Therefore, we conclude that AP-PCR

DNA fingerprinting is important and practically feasible technique for elucidating the genetic background of various tumors. Accordingly, we believe that this technique is rather neglected in contemporary research and should make a comeback because it still has a particularly promising future in experimental oncology.

## 5. Acknowledgments

This study was supported by Grant # III41031 from the Ministry of Education and Science, Republic of Serbia.

## 6. References

Bankovic, J.; Stojsic, J.; Jovanovic, D.; Andjelkovic, T.; Milinkovic, V.; Ruzdijic, S. & Tanic N. (2010). Identification of genes associated with non-small-cell lung cancer promotion and progression. *Lung Cancer*, Vol.67, No.2, pp. 151-159, ISSN 0169-5002.

Beivik, J. 2001. Don't stop for repairs in a war zone: Darwinian evolution unites genes and environment in cancer development. *PNAS*, Vol.98, No.10, pp. 5379-5381, ISSN 0027-8424.

Breivik, J. & Gaudernack G. (1999). Genomic instability, DNA methylation and natural selection in colorectal carcinogenesis. *Seminars in Cancer Biology*, Vol.9, No.3, pp. 245-254, ISSN 1044-579X

Cahill, D.P.; Kinzler, K.W.; Vogelstein, B, & Lengauer, C. (1999) Genetic instability and Darwinian selection in tumours. *Trends Cell Biology*, Vol.9, No.12, pp. M57–M60, ISSN 0962-8924.

Chariyalertsak, S.; Khuhaprema, T.; Bhudisawasdi, B.S.; Wongkham, S. & Petmitr S. (2005). Novel DNA amplification on chromosome 2p25.3 and 7q11.23 in cholangiocarcinoma identified by arbitrarily primed polymerase chain reaction. *Journal of Cancer Research & Clinical Oncology*, Vol.131, No.12, pp. 821–828, ISSN 1432-1335.

Cobb, B. (1997) Optimization of RAPD fingerprinting. In: *Fingerprinting Methods Based on Arbitrarily Primed PCR*, Micheli M.R. & Bova R., (ed), pp. 93-102, Springer Lab Manual, Springer-Verla, ISBN 3-540-61229-7, Berlin, Germany.

Dawkins, R. (1989). *The selfish gene* (edition 1). Oxford University Press, ISBN 0192860925, Oxford, UK.

Eigen, M. (1993). The origin of genetic information: viruses as models. *Gene*, Vol.135, No.1-2, pp. 37–47, ISSN 0378-1119.

Fenech, M. (2002). Chromosomal biomarkers of genomic instability relevant to cancer. *Drug Discovery Today*, Vol.7, No. 22, 1128–1137, ISSN 1359-6446.

Hunter, K.D.; Parkinson, E.K. & Harrison, P.R. (2005). Profiling early head and neck cancer. *Nature Reviews Cancer*, Vol.5, No pp. 127–135, ISSN: 0028-0836.

Jass, J.R.; Mukawa, K.; Goh, H.S. & Love, S.B. (1989) Clinical importance of DNA content in rectal cancer measured by flow cytometry. *Journal of Clinical Pathology*, Vol.42, No.3, pp. 254–259, ISSN 0021-9746.

Kallioniemi,O.P.; Kallioniemi, A.; Piper, J.; Isola, J.;Waldman, F.M.; Gray, J.W. & Pinkel, D. (1994). Optimizing comparative genomic hybridization for analysis of DNA sequence copy number changes in solid tumors. Genes Chromosomes Cancer, Vol.10, No.4, pp. 231–243, ISSN 1098-2264.

Lengauer, C.; Kinzler, K.W. & Vogelstein, B. (1998). Genetic instabilities in human cancers. *Nature*, Vol.396, pp. 643–649, ISSN 0028-0836.

Loeb, L.A. (2001). A mutator phenotype in cancer. *Cancer Research*, Vol.61, pp. 3230–3239, ISSN: 0008-5472.

Loeb, L.A. (1991). Mutator phenotype may be required for multi-stage carcinogenesis. *Cancer Research*, Vol.51, pp. 3075–3079, ISSN: 0008-5472.

Markovic, J.; Stojsic, J.; Zunic, S.; Ruzdijic, S. & Tanic, N. (2008) Genomic instability in patients with non-small cell lung cancer assessed by the arbitrarily primed polymerase chain reaction. *Cancer Invest*, Vol.26, No. pp. 262-8, ISSN: 0735-7907.

McClelland, M. & Welsh, J. (1994). DNA fingerprinting by arbitrarily primed PCR. *PCR Methods Application*, Vol.4, pp. S59-S65.

Misra, A.; Chattopadhyay, P.; Chosdol, K.; Sarkar, C.; Mahapatra, A.K.; & Sinha S. (2007). Clonal mutations in primary human glial tumors: evidence in support of the mutator hypothesis. *BMC Cancer*, Vol.7, pp. 190. ISSN 1471-2407.

Negrini, S.; Gorgoulis, V.G. & Halazonetis, T.D. (2010) Genomic instability-an evolving hallmark of cancer. *Nat Rev Mol Cell Biol*, Vol.11, pp. 220-228, ISSN 1471-0072.

Neville, B.W. & Day, T.A. (2002). Oral cancer and precancerous lesions. *CA Cancer J Clin*, Vol.52, No.4, pp. 195–215, ISSN 0007-9235.

Nishizaki, T.; Harada, K.; Kubota, H.; Harada, K.; Furuya, T.; Suzuki, M. & Sasaki K. (2002). Chromosome instability in malignant astrocytic tumors detected by fluorescence in situ hybridization. *Journal of Neurooncology*, Vol.56, No.2, pp. 159-165, ISSN 0167-594X.

Ohgaki, H. & Kleihues, P. (2007). Genetic pathways to primary and secondary glioblastoma. *Am J Pathol*, Vol.170, No.5, pp. 1445-53, ISSN 0002-9440.

Ong, T.M.; Song, B.; Qian, H.W.; Wu, Z.L. & Whong, W.Z. (1998). Detection of genomic instability in lung cancer tissues by random amplified polymorphic DNA analysis. *Carcinogenesis*, Vol.19, No.1, pp. 233-235, ISSN 0143-3334.

Peinado, M.A.; Malkhosyan, S.; Velazquez, A.; & Perucho, M. (1992). Isolation and characterization of allelic losses and gains in colorectal tumors by arbitrarily primed polymerase chain reaction. *Proc. Natl. Acad. Sci. USA*, Vol.89, No.21, pp. 10065–10069, ISSN 0027-8424.

Perucho, M. (1996). Microsatellite instability: the mutator that mutates the other mutator. *Nat. Med.*, Vol.2, No.6, pp. 630–631, ISSN 1078-8956.

Sambrook, J.; Fritch, E.F. & Maniatis, T. (1989). *Molecular cloning: a laboratory manual.* (2nd edition). Cold Spring Harbor Laboratory Press, Cold Spring Harbor, NY.

Tanic, N.; Tanic, N.; Milasin, J.; Vukadinovic, M. & Dimitrijevic, B. (2009). Genomic instability and tumor-specific DNA alterations in oral leukoplakias. *European Journal of Oral Sciences*, Vol.117, pp. 231 – 237, ISSN 0909-8836.

Tomlinson, I.P.; Novelli, M.R.; Bodmer, W.F. (1996). The mutation rate and caner. *Proc. Natl. Acad. Sci. USA*, Vol.93, No.25, pp. 14800–14803, ISSN 0027-8424.

Van Meir, E.G.; Hadjipanayis, C.G.; Norden, A.D.; Shu, H.K.; Wen, P.Y.; & Olson, J.J.(2010). Exciting new advances in neuro-oncology: the avenue to a cure for malignant glioma. CA Cancer J Clin, Vol.60, No.3, pp. 166-93, ISSN 0007-9235.

Vogelstein, B.; Fearon, E.R.; Kern, S.E.; Hamilton, S.R.; Preisinger, A.C.; Nakamura, Y. & White, R. (1989). Allelotype of colorectal carcinomas. *Science*, Vol.244, No.4901, pp. 207–211, ISSN: 0036-8075.

Wang, J.; Wang, Q. & Ye, F. (2002). Genetic instability in cancer tissues analyzed by random amplified polymorphic DNA PCR. *Chin Med J (Engl)*, Vol.115, No.3, pp. 430-432, ISSN 0366-6999.

Welsh, J. & McClelland, M. (1990). Fingerprinting genomes using PCR with arbitrary primers. *Nucleic Acid Research*, Vol.18, No.24, pp. 7213–7218, ISSN 0305-1048.

Williams, J.G.K.; Kubelik, A.R.; Livak, K.J.; Rafalski, J.A. & Tingey, S.V. (1990). DNA polymorphisms amplified by arbitrary primers are useful as genetic markers. *Nucleic Asids Research*, Vol.18, No.22, pp. 6531-6535, ISSN 0305-1048.

# 5

# Analysis of Alternatively Spliced Domains in Multimodular Gene Products - The Extracellular Matrix Glycoprotein Tenascin C

Ursula Theocharidis and Andreas Faissner
*Ruhr-University Bochum*
*Germany*

## 1. Introduction

In 1977 it was discovered that the one-gene-one-enzyme hypothesis was not true (Chow et al.,1977; Berget at al., 1977). The primary transcription product can be spliced in different ways and give rise to several proteins depending on the exons being present in the final mRNA. This phenomenon is called alternative splicing and indeed is common to many genes. Several possible modes of alternative splicing are known and the most common one is the inclusion or exclusion of an exon, the exon skipping.

Based on polymerase chain reaction (PCR) techniques we developed a method to analyse combinations of alternatively spliced domains in multimodular gene products. This method was used to determine the combinatorial variability of tenascin C isoforms in the mouse central nervous system (Joester & Faissner, 1999) and in neural stem cells (von Holst et al., 2007).

Here, we present the method of amplifying different sized isoforms of a gene product with several alternatively spliced domains via PCR and the isolation and subcloning of the PCR products. Clones are analyzed for alternatively spliced domains contained therein by a dot blot *in vitro* hybridization method with domain-specific DNA probes which were generated using PCR.

## 2. Background information

Tenascin C is a multimodular glycoprotein of the extracellular matrix which is mainly expressed during central nervous system development and in pathological states such as brain tumours or lesions. We have studied the expression pattern of this molecule and its function *in vivo* and *in vitro* and collected evidence concerning its structural diversity. We and others determined its functions during neural development, in the adult neural stem cell niche and in lesions and tumours (Czopka et al., 2009, 2010; Dobbertin et al., 2010; Garcion et al., 2001, 2004; Garwood et al., 2011; Gates et al., 1995; Orend & Chiquet-Ehrismann, 2006; von Holst et al., 2007).

Tenascin C contains a constant part including eight contitutive fibronectin type III (fnIII) domains and a variable part of six alternatively spliced fnIII domains in the mouse which can be included independently into the gene product (figure 1).

The alternatively spliced fnIII domains of the tenascin C molecule have different functions, e.g. affecting the axon outgrowth of developing nerve cells or the migration potential of brain tumour cells (Rigato et al., 2002; Michele & Faissner, 2009; Broesicke & Faissner, personal communication). Therefore it is important to have a method to determine the isoform composition of the molecule in the tissue or cell cultures used.

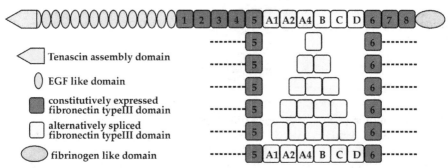

Fig. 1. Schematic representation of mouse tenascin C. The monomer consists of several distinct protein domains. At the N-terminal tenascin C assembly domain six monomers can be assembled to the so called hexabrachion (Erickson & Inglesias, 1984). 14,5 epidermal growth factor (EGF) like domains and eight constitutive fibronectin type III (fnIII) domains follow before the C-terminal globular lobe homologous to the beta- and gamma-chains of fibrinogen. Between the fifth and sixth constant fnIII domain up to six alternatively spliced domains can be inserted and an independent alternative splicing at each position could lead to the generation of 64 (=$2^6$) possible isoforms of the molecule. All possible numbers of domains can be inserted in the final splicing product, but the combination of cassettes is unclear in most cases. Only the largest variant necessarily contains all six alternatively spliced domains.

Gene products with different exons being alternatively spliced and inserted into the sequence can be distinguished by PCR when the sizes of the resulting mRNAs are different. A PCR analysis uses primers flanking the alternatively spliced region and results in amplicons with different sizes. These can be analysed by agarose gel electrophoresis and show bands in distinct positions. Tenascin C has six domains that can be alternatively spliced and independently inserted into the sequence. The analysis of these domains on an agarose gel shows the size of the resulting amplicons and therefore the number of inserted domains but leaves the question open which of the possible domains are included. A further analysis is therefore needed. We have shown that it can be performed using an *in vitro* dot blot hybridization technique to verify the exact domain combinations.

## 3. Analysis of isoform sizes in multimodular gene products by RT-PCR

When the sequence of the gene of interest is known primers can be generated which allow the amplification of the relevant region. The primers can either bind in the alternatively spliced region itself and therefore generate PCR products only when the target sequence is expressed. When the primers bind outside of the alternatively spliced part of the sequence the products can contain every possible insert additionally to the constant parts of the sequence which are defined by the primer binding sites. Additionally, isoforms without any insert can be amplified with these primer combinations.

Tenascin C has its variable region between the constantly expressed fnIII domains 5 and 6. We used two different primer combinations to determine the isoform pattern of the molecule in various tissues and cell cultures (figure 2). The primers 5s and 6as bind to the 5'end of the fifth and the 3'end of the sixth domain and result in PCR products with the smallest form containing only these two constant domains. Another primer pair we used was called 5for and 6rev and these bind to the 3'end of domain number 5 and the 5'end of domain number 6. The smallest amplicons are then represented by forms with one alternatively spliced fnIII domain. The further analysis was carried out with PCR products obtained with this primer pair.

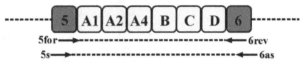

Fig. 2. Primer binding sites. Two different primer pairs were used to amplify the alternatively spliced region of tenascin C and analyse the expression profile of different isoforms. The primers 5s / 6as and 5for / 6rev bind to the constant fnIII domains 5 and 6 at their outer or inner tails, respectively. The primers 5s and 6as bind to the 5'end of the fifth and the 3'end of the sixth domain. The resulting amplicons therefore contain the minimum of two fnIII domains, namely 5 and 6. The further insertion of alternatively spliced cassettes increases the size of the PCR product. The primers 5for and 6rev bind to the 3' end of fnIII domain 5 and the 5'end of fnIII domain 6. Only PCR products with the minimum of one alternatively spliced fnIII domain can be generated. Every additional domain increases the size of the amplicons by 273 bp, the size of the single domains.

### 3.1 Expression analysis by RT-PCR

The expression analysis can be performed on RNA isolated from tissue or cell culture material which was processed by reverse transcription. Several commercially available kits help to isolate total RNA or mRNA from tissue or cell cultures. The resulting RNA can then be used to generate cDNA by reverse transcription which can also be carried out using kits from different suppliers. If oligo-dT primers or random primers are used for the reverse transcription makes no difference in our experience. The generated cDNA is the template for the PCR which possibly needs some optimization steps to generate all bands of interest. According to our experience it is of outstanding importance to test the performance of different Taq polymerases in advance because not every enzyme from each supplier will work equally efficiently. Different polymerases in their respective buffer system show variable results and should be adapted to the reaction requirements.

The PCR conditions with regard to annealing temperature and time, elongation time as well as concentration of cDNA, primers and Magnesium must be worked out in advance. Addition of DMSO or betain may be needed and checked when the standard conditions don't lead to the desired results. To be able to generate all the expected amplicons the longest product determines the elongation time. The rule of thumb to calculate 1 minute elongation time per 1000 base pairs gives a good estimation here.

The resulting PCR products can be processed on an agarose gel and the DNA bands made visible with ethidium bromide or a substitute. The concentration of the agarose must be high enough to discriminate between contiguous bands but sufficiently low that the longest products can enter the analysis area. A long gel chamber increases the migration way and a lower voltage over a longer time period narrows the single bands and makes the discrimination easier.

We used brain tissue from postnatal mice or cultures of neural stem cells to isolate total RNA and analysed the expression pattern of the alternatively spliced forms of Tenascin C in the respective system (Joester & Faissner, 1999; von Holst et al., 2007). In these cases we found isoforms of all possible sizes to be present and performed the further analysis for isoforms containing between one and six additional cassettes. The use of the primer pair 5s and 6as leads to DNA bands on the agarose gel where the smallest one is 546bp, representing only the two constant fnIII domains 5 and 6 of 273bp each. Every additional cassette increases the amplicon size by 273bp. Therefore, we can see a "ladder" structure of up to seven DNA bands on the agarose gel when using this primer pair (figure 3B). When the primers 5for and 6rev are present in the PCR mix instead we get products where the smallest isoform contains the minimum of one alternatively spliced fnIII domain. The larger bands represent the larger forms with up to six fnIII domains. Here, we get the maximum of 6 DNA bands on the gel (figure 3A).

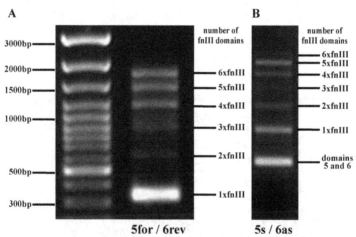

Fig. 3. Examples of tenascin C isoform PCRs. The primer pairs 5for / 6rev and 5s / 6as were used to amplify the alternatively spliced region of tenascin C. The PCR products were separated on an 1,2 % agarose gel. (A) The smallest amplicon generated using the primers 5for and 6rev contains only one of the alternatively spliced cassettes. The insertion of additional domains increases the product size by 273 bp. Up to six bands appear representing the different possible amplicon sizes. PCR products amplified with this primer pair were used for the further analysis of the domain expression profile after separation on an agarose gel. (B) The use of the alternative primer pair 5s and 6as leads to the generation of up to seven DNA bands on the agarose gel because the smallest band represents only the constant fnIII domains 5 and 6 without any insert. When alternatively spliced domains are included in the sequence the product size increases by 273 bp for each domain. Up to six domains can be added and therefore the largest DNA band on the gel represents the total of eight fnIII domains. This primer pair was mainly used for the analysis of expression profiles.

The agarose gel shows the expression profile of the alternatively spliced gene products in the analysed tissue or cells. The resulting amplicons answer the first questions in this respect: Are different forms expressed in parallel? Are all possible product sizes present? What is the ratio between different forms? Does the expression profile change with the conditions?

## 3.2 Cloning of resulting PCR products

The analysis of the PCR products on an agarose gel answers the question for size and ratio of the isoforms expressed but leaves open which of the possible domains are contained in the bands. Some further experimental steps are necessary to determine the domains being expressed. Because several domain combinations can migrate in the same position they must be separated from each other. This can be achieved by subcloning the different PCR amplicons and analysis of the resulting clones.

The PCR bands are cut out of the gel under visual control at an UV desk and the gel slices collected in separate tubes. It is important to use different knives for each band because otherwise DNA from other bands might be carried over and contaminate the samples. Isolation of the DNA from the gel can be performed using classic methods or commercially available kits. The elution should be done with the minimal amount of water to avoid problems with following reaction steps. For the subcloning of the PCR products we used the TOPO-TA cloning kit from invitrogen but any other similar kit will do. In our experience it is important to handle the bacteria quite carefully and leave them grow in antibiotic-free medium for 30 minutes after the transformation. Spread the bacteria to LB agar plates with appropriate antibiotics then and let them grow over night.

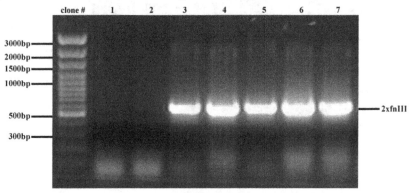

Fig. 4. Check for positive clones after direct colony lysis. The colonies grown after the cloning and transformation of the PCR products are checked for their content of fnIII domains. The primer pair 5for / 6rev was used to generate amplicons of the expected size when the clones have taken up the plasmids containing the fnIII domains. This example shows seven clones from a DNA band containing 2 fnIII domains. Two of the clones shown here do not contain any fnIII domains and are therefore not selected for the following screen. The other clones show PCR bands of the expected size and are analysed in the subsequent dot blot hybridizations.

The content of the resulting clones can quickly and easily be checked by direct lysis of the bacteria and a subsequent PCR with the primers used before. The colonies grown on the agar plate are picked with a pipette tip and transferred to another (the "master" plate) into numbered fields. The tip is then shaken in $25\mu$L 70% ethanol in PCR tubes to lyse the bacteria. The master plate can be placed in the incubator while in the meantime the colonies are checked for their content. In an incubation step the ethanol is evaporated at $80°$C for approx. 15

minutes before the PCR master mix containing buffer, primers and polymerase is added. The reaction conditions can be the same as before. The products can be analysed on an agarose gel and should show single bands in the expected position for each positive clone (figure 4). These can subsequently be picked from the master plate and propagated in miniprep scale. The plasmid DNA from the miniprep cultures can be isolated by alcaline lysis or with appropriate kits.

### 3.3 Analysis of clones - dot blot

Of course, these plasmids could be sequenced and their composition clarified by this method at this point. Because sequencing is not cheap when analysing hundreds of clones a method was developed that renders the identification of many samples in one step possible and is cheaper. The basis is a dot blot of the isolated plasmids to nylon membranes which subsequently can be used in hybridizations with domain specific probes.

The plasmid solutions should be adjusted to similar concentrations with water to have equal amounts of target DNA in the spots. The easiest way to apply the plasmids to the membranes is the use of a dot blot apparatus with a vacuum manifold, but it is also possible to spot the liquid using a master plate onto the membrane which is placed on filter paper. For our analysis we used Hybond N+ membranes from Amersham which were pre-wetted with 10x SSC (1,5 M NaCl; 150 mM Na$_3$Citrate, pH 7,0) buffer. Because there are six possible domains to detect (A1, A2, A4, B, C and D) and we used a negative control we prepared seven membranes with identical spot patterns. The plasmid solutions were diluted in 10x SSC in a volume of 100$\mu$L when we used the dot blot apparatus and 6$\mu$L when a pattern was used.

After application of the plasmid solutions the membranes are incubated for 10 minutes in denaturing buffer (500 mM NaOH; 1,5 M NaCl) and 10 minutes in renaturing buffer (500 mM Tris/HCl, pH 7,5; 1,5 M NaCl) to prepare the DNA for the hybridization. The nylon membranes are dried and baked at 80°C for two hours to have the DNA bound covalently to them. These dot blots can be stored for some time at room temperature.

### 3.4 Positive and negative controls

To determine the specificity of the method and to be sure that no false-positive or false-negative results appear the use of positive and negative controls is important. For every application appropriate controls must be defined. In our case we could exploit the fact that the fnIII domain number 6 is not included in the alternatively spliced region which we amplify with the primer pair 5for / 6rev in the initial PCRs. Therefore a probe detecting the fnIII domain 6 serves as negative control. Another control we use is a plasmid, called pJT1# which contains the constant part of tenascin C between the fnIII domains 2 and 8, but none of the alternatively spliced domains. The positive control is a plasmid containing all six alternatively spliced domains. On the dot blot it is applied in addition to the clones under investigation.

### 3.5 Generation of domain specific DNA probes by PCR

The hybridization of membrane-bound DNA with probes detecting defined DNA fragments identifies specific sequences in the bound nucleic acids. Probes detecting the desired target sequences are generated based on the cDNA of these fragments which are cloned into common plasmid vectors. We used the sequences of the tenascin C fnIII domains A1, A2, A4, B, C, D and 6 as negative control which were inserted in pBluescript II KS+ vectors.

These inserts are labelled to use them in expression studies. The labelling with fluorescein has several advantages. When using non-radioactive probes no special safety regulations must be obeyed. Additionally, the probes can be used for a longer time period. This is of special advantage when several probes are used in parallel. Manufacturer's data state that fluorescein-labelled probes are stable for 6 months without decreasing activity. Radioactively labelled probes would lose sensitivity after a few days because the isotopes disintegrate continuously. Indeed, the probes generated in our lab could be used for several years (Joester & Faissner, 1999; von Holst et al., 2007).

The labelling was performed with fluorescein-coupled dUTP (Amersham). The manufacturer's labelling kit could not be used because it is based on a random primer labelling method. This uses the hybridization of 8 to 10 bases long random primers to single DNA strands. In a polymerization mix with the labelled nucleotide the Klenow fragment of the DNA polymerase I generates the complementary probe. But the tenascin C domains are less than 300 base pairs long and therefore have only a few potential binding sites for random primers which may lead to only very short probes. The labelling efficiency is too low (1 labelled base in 50 bases, according to manufacturer's data) to achieve an appropriate labelling frequency. Therefore we developed a labelling protocol which uses a PCR method to generate dUTP-labelled DNA probes.

The Taq polymerase incorporates dUTP with less efficiency than unlabelled dTTP. Therefore the exclusive use of dUTP would show the optimum of labelled probe but only low yield of amplification product. A low amount of fluorescein-dUTP and higher amount of dTTP reverses this effect and leads to a higher product yield but low labelling efficiency. We adjusted the PCR conditions to the optimal yield of labelled amplification product.

The optimal reaction conditions required only 10 pg of plasmid DNA and a low amount of dNTPs (20 $\mu$M). The reaction mix contained 60 mM Tris/HCl, pH 8,5; 15 mM $(NH_4)_2SO_4$; 2 mM $MgCl_2$; 0,2 $\mu$M sense primer; 0,2 $\mu$M antisense primer and 1 Unit Taq polymerase in 25 $\mu$L volume. The cycling conditions are dependent on the hybridization temperature of the respective primers and the length of the expected product.

In the first labelling reactions with domain C different amounts of dTTP were replaced with Fl-dUTP (3 to 50% equivalent to 0,6 to 10 $\mu$M Fl-dUTP). The amount of the products increases with decreasing amounts of the labelled nucleotide. The labelling efficiency was also tested on dot blots with different concentrations of the plasmid containing the C domain. 1$\mu$L of the PCR products were used in the hybridization solution. After an over-night incubation the blots were developed with an alkaline phosphatase-coupled antibody detecting fluorescein. The detection sensitivity was proportional to the concentration of the Fl-dUTP used in the labelling reaction. The subsequent labelling reactions were performed using 17,5 $\mu$M dTTP and 2,5 $\mu$M fluorescein-11-dUTP. All probes detecting the fnIII domains A1, A2, A4, B, C and D of tenascin C were labelled with this method and called Fl-A1, Fl-A2,... Figure 5 shows the resulting PCR amplicons.

The fluorescein-labelled probes were tested for their detection capability of different dilutions of the respective plasmids. The senstivity was different for the probes and therefore their concentration was adjusted in the hybridization solution. The hybridization results show that the sensitivity is equal between 3 pg up to 1 ng of the target sequence (figure 6A). This sensitivity is much higher than that seen for agarose gels stained with ethidium bromide which is in the range of 1 ng DNA. The sensitivity was tested regularly to adjust the stability or labelling efficiency of the different probes but none showed a significant reduction in detection efficiency over time.

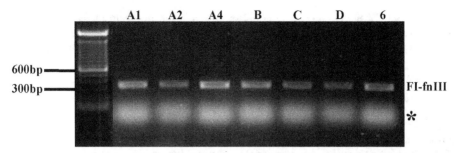

Fig. 5. Fluorescein-labelled probes on an agarose gel. The DNA probes labelled with FI-dUTP
were applied to an agarose gel and show bands in the expected size of less than 300 bp. The
asterisk designates the fluorescence signal of the non-incorporated nucleotides.

The specificity of the fluorescein-labelled probes was tested with seven dot blot stripes
containing the plasmids pA1, pA2, pA4, pB, pC, pD, p6 and pJT1# in equal concentrations.
The stripes were hybridized with the probes for the single fnIII domains (FI-A1, FI-A2,...) and
the alkaline phosphatase reaction was developed (figure 6B). Highly stringent hybridization
and washing conditions minimised the cross-reactivity of the probes with unspecific target
sequences. These conditions included the hybridization and the first washing steps at 72°C

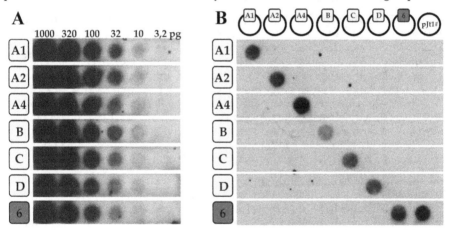

Fig. 6. (A) Sensitivity of fluorescein-labelled probes. The plasmids containing the single fnIII
domains A1, A2, A4, B, C, D and 6 (designated pA1, pA2, ...) were diluted and applied to the
nylon membranes in dots. The membranes were incubated with the respective probes and
the reaction developed with an anti-fluorescein antibody coupled to alkaline phosphatase. A
minimum of about 10 pg of target sequence was detected by each of the probes. The probes
were diluted so that all of them detected their targets in a comparable way. (B) Specificity of
fluorescein-labelled probes. Seven identical dot blots containing 10 ng of the plasmids pA1,
pA2, pA4, pB, pC, pD, p6 and 19 ng pJT1# (corresponding to 1 ng target sequence) were
hybridized with the different fluorescein-labelled probes. The probes detect their target
sequences with high specificity. Although domains A1 and A4 are highly identical the probes
do not show a significant cross-reactivity. The probe FI-6 detects the plasmid pJT1# which
contains domain number 6.

and the use of 0,5% SDS in the washing buffer. The highest probability for a cross-reactivity exists between the domains A1 and A4 because their nucleotide sequence is 80% identical. Only in a few cases a light background signal could be detected when using these probes.

## 3.6 Hybridization

The generated probes are used in an *in vitro* hybridization protocol and applied to the nylon membranes containing the plasmid DNA from the clones which shall be analysed. The nylon membranes with the bound plasmids are washed in 5xSSC and pre-incubated in hybridization solution (5x SSC; 0,1% SDS; 5% dextrane sulfate; 5% liquid block (Amersham)) for 30 minutes at 72°C with gentle agitation. An appropriate amount of the probes which must be determined in preliminary experiments is added to 200$\mu$L of hybridization solution. The probes are denatured at 96°C for 5 minutes and applied to the membranes. The hybridization takes place over night at 72°C with constant agitation in a hybridization oven.

## 3.7 Signal detection

We used two different methods for the detection of the hybridized probes. A detection protocol to obtain chemoluminescence signals uses the fluorescein gene images CDP-*Star* detection system (Amersham). The other option was the development of a colour reaction using NBT and BCIP as alkaline phosphatase substrates.

When the DNA on the membranes was hybridized over night with the fluorescein-coupled probes the membranes can be washed for 2 x 15 minutes in wash buffer 1 (0,1x SSC; 0,5% SDS) at 72°C. To block unspecific binding sites they are incubated for one hour in blocking buffer (10% liquid block (Amersham) in detection buffer (100 mM Tris/HCl, pH 7,5; 300 mM NaCl)) before the alkaline phosphatase-coupled anti-fluorescein antibody (1:5000 in detection buffer with 0,5% BSA) is applied for an hour. Unbound antibody is washed away with wash buffer 2 (0,3 % Tween-20 in detection buffer) 4x 8 minutes. For the chemoluminescence detection the blot is moistened with a dioxetane-based substrate solution (CDP-*Star* detection reagent (Amersham). After 3 minutes the excess substrate solution is removed and the blot placed between two sheets of foil and laid on an autoradiographic film. Depending on the DNA concentration the optimal detection time was between 10 and 60 minutes.

For the alternative developing method the membranes are washed 3x 5 minutes in wash buffer A (100 mM Tris/HCl, pH 7,4; 150 mM NaCl; 0,3% Tween-20) after the antibody incubation, 2x 5 minutes in wash buffer B (100 mM Tris/HCl, pH 9,5; 100 mM NaCl) and 3x 10 minutes in TBS (50 mM Tris/HCl, pH 7,5; 150 mM NaCl). To develop the colour reaction the membranes are wetted with staining solution containing NBT and BCIP (Roche) and not shaken any more. The colour reaction will appear after five to 60 minutes. The reaction can be stopped with water and the membranes dried afterwards. Because the detection sensitivity is lower for the colour reaction a higher amount of plasmid DNA must be used in this case. 50 ng of target sequence in the spots lead to good results (data not shown).

## 3.8 Analysis of domain combinations

The read-out of the results is straight forward. The membranes show positive signals whenever the respective domain is present in the clone. Every plasmid DNA shows a specific pattern of positive and negative signals and therefore stands for the presence or absence of a given single domain.

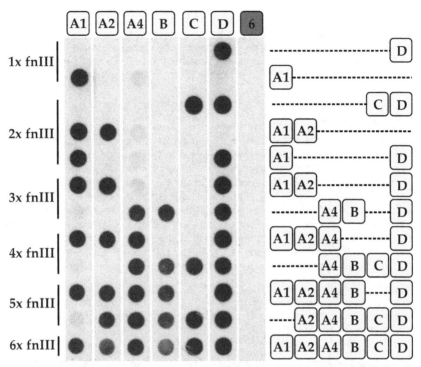

Fig. 7. Example of screening results. Plasmids containing different numbers of alternatively spliced fnIII domains were applied to nylon membranes in dots. Seven identical blots were generated and hybridized with the fluorescein-labelled probes FI-A1, FI-A2, FI-A4, FI-B, FI-C, FI-D and FI-6 as negative control. After the development of the alkaline phosphatase reaction the dot blots show positive signals whenever the respective fnIII domain is present in the plasmid. Therefore the domain combination of every single clone can be directly read out from the blots.

In intensive studies of the expression pattern of Tenascin C isoforms in the developing brain and in neural stem cells (Joester & Faissner, 1999 and von Holst et al., 2007) we detected 28 different isoforms of Tenascin C out of 64 possible ones which could theoretically be generated with six independently spliced domains ($=2^6$). We had several hundred clones to identify which contained between one and six alternatively spliced fnIII domains. The membranes we prepared were handled separately depending on the expected number of domains to be present in the plasmids. Figure 7 shows an example of the analysis of several clones with different numbers of fnIII domains. Plasmids from the distinct subcloning reactions were spotted onto seven nylon membranes and hybridized with the probes FI-A1, FI-A2, FI-A4, FI-B, FI-C, FI-D and FI-6 as negative control. The signals show that different combinations of fnIII domains can be contained in the plasmids. The clones with only one alternatively spliced domain displayed here for example contain the domains A1 or D, respectively. Indeed, these were the most common domains among single-domain clones when a complete screen was performed (Joester & Faissner, 1999; von Holst et al., 2007). The variability of domain combinations is higher in the middle-size clones with two, three or four alternatively spliced fnIII domains. The plasmids containing five additional cassettes on the other hand show usually the absence of fnIII domain C and only few miss A1. The extensive screens for the expression profiles of fnIII domains being expressed in postnatal mouse cerebellum or neural stem cells show that the possible variability among the clones is not utilised. 64 different isoforms of tenascin C would be theoretically possible but only 28 forms were found. Some combinations of domains were never seen like the direct link between the fnIII domains C and 6 or A4 and C.

### 3.9 PCR of single domains

To confirm the results of the hybridization and to clarify ambiguous signals we carried out PCRs for the single fnIII domains that were detected in the plasmid DNA. It is important to highly dilute the plasmid DNA and use only 10 to 20 pg plasmid DNA as template and to use highly specific PCR conditions. The specificity of the PCR conditions was confirmed before because the fnIII domains show high similarities and could therefore lead to false-positive signals when using standard PCR conditions on plasmid templates. We used a 2-step PCR with a high annealing temperature of the primers of 72°C and combined the annealing step with the elongation step to a 40-seconds 72°C incubation step. The cycling was therefore between 20 seconds 94°C and 40 seconds 72°C. We also skipped the final 5-minutes elongation step which we usually applied, especially for the addition of adenosines for cloning purposes.

Figure 8 shows the high specificity of these PCR conditions when the different fnIII domain-containing plasmids were used in these reactions. Only for those plasmids amplicons were generated when the respective primer pair was used. Therefore, we had an additional tool to confirm the dot blot results.

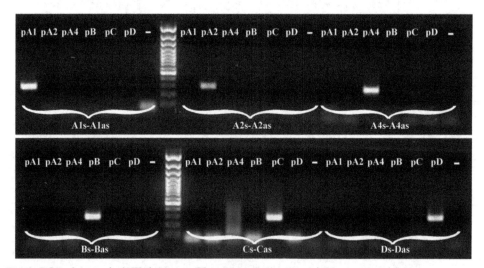

Fig. 8. PCRs for single fnIII domains. The primer pairs A1-s / A1-as, A2-s / A2-as,... were used in PCRs for the amplification of single fnIII domains. The domains have some similarities in their sequences. Therefore the PCR conditions must be highly specific. PCRs with stringent conditions amplify only products from plasmids containing the respective domain. Such conditions can be used to test plasmids with unclear domain composition.

Analysis of Alternatively Spliced Domains in Multimodular Gene Products - The Extracellular
Matrix Glycoprotein Tenascin C

69

## 4. Adaptation to general application

Many genes are subject to alternative splicing and most of them show an exon skipping mode
which implies the inclusion or exclusion of single exons. When the possible exon structure
leading to the appearance or absence of single domains is known the expression profile of

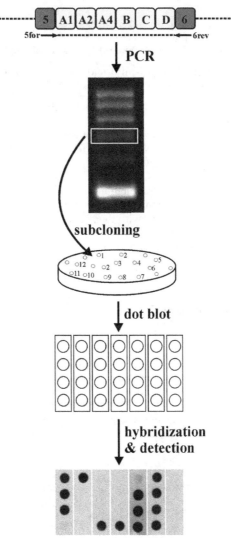

Fig. 9. Schematic presentation of the method. Primers flanking the alternatively spliced
region of the molecule are used to generate PCR products of different sizes which are
separated on an agarose gel. The single bands are cut out off the gel and subcloned
separately. The resulting clones can be analysed with a dot blot hybridization procedure with
non-radioactively labelled probes. Positive and negative signals display the domain
composition of every single clone.

the domains can be analysed using the method presented here. Some preliminary steps are necessary before a screen for expressed domains can be started but when the system is set up once it can be used for the screening of many PCR products over a long time.

To start such a screen the following steps must be accomplished:
1. Clarify the domain sequence
2. Generate primers detecting the single domains
3. Clone the single domains into plasmid vectors
4. Use these vectors as templates for PCRs generating dUTP-labelled DNA probes

A screen includes the generation of the plasmids and the dot blot before the hybridization can start. Therefore conduct the following steps to start a screen:

1. Isolate RNA from the tissue or cell type under investigation
2. Prepare cDNA based on this RNA
3. Use this cDNA as template in PCRs for the alternatively spliced region of your gene
4. Separate the amplicons on an agarose gel and cut off the single bands
5. Clone the PCR products into plasmid vectors
6. Dilute the plasmid vectors to appropriate concentrations
7. Apply the plasmid solutions onto nylon membranes
8. Denature the DNA and bind it covalently to the membranes
9. Hybridize the DNA on the membranes with the probes
10. Wash under stringent conditions
11. Apply an antibody to the labelling marker
12. Develop the enzyme reaction
13. Read out your domain structure

Figure 9 shows the summary of the method:

## 5. Conclusion

With the method presented here we developed a possibility to unravel unknown structures of splice products for alternatively spliced transcripts. The example we analysed was the extracellular matrix molecule tenascin C but any other multimodular protein can be examined in a similar way. With some preliminary preparations an operational tool is at hand which makes the screening of many clones and therefore the generation of an expression profile possible.

## 6. Acknowledgements

The authors wish to thank very much Dr. Angret Joester for providing unpublished material concerning the dot blot assays. We also thank the German Research Foundation (DFG) (SPP 1048, Fa 159/11-1, 2, 3) and the GRK 736 for support.

## 7. References

Berget, S.M., Moore, C. & Sharp, P.A. (1977) Spliced segments at the 5' terminus of adenovirus 2 late mRNA. Proc Natl Acad Sci U S A. 1977 Aug;74(8):3171-5.
Chow, L.T., Gelinas, R.E., Broker, T.R. & Roberts, R.J. (1977) An amazing sequence arrangement at the 5' ends of adenovirus 2 messenger RNA. Cell. 1977 Sep;12(1):1-8.

Czopka, T., Von Holst, A., Schmidt, G., Ffrench-Constant, C. & Faissner, A. (2009) Tenascin C and tenascin R similarly prevent the formation of myelin membranes in a RhoA-dependent manner, but antagonistically regulate the expression of myelin basic protein via a separate pathway. Glia. 2009 Dec;57(16):1790-801.

Czopka, T., von Holst, A., ffrench-Constant, C. & Faissner, A. (2010) Regulatory mechanisms that mediate tenascin C-dependent inhibition of oligodendrocyte precursor differentiation. J Neurosci. 2010 Sep 15;30(37):12310-22.

Dobbertin, A., Czvitkovich, S., Theocharidis, U., Garwood, J., Andrews, M.R., Properzi, F., Lin, R., Fawcett, J.W. & Faissner, A. (2010) Analysis of combinatorial variability reveals selective accumulation of the fibronectin type III domains B and D of tenascin-C in injured brain. Exp Neurol. 2010 Sep;225(1):60-73. Epub 2010 May 5.

Erickson, H.P. & Inglesias, J.L. (1984) A six-armed oligomer isolated from cell surface fibronectin preparations. Nature. 1984 Sep 20-26;311(5983):267-9.

Garcion, E., Faissner, A. & ffrench-Constant, C. (2001) Knockout mice reveal a contribution of the extracellular matrix molecule tenascin-C to neural precursor proliferation and migration. Development. 2001 Jul;128(13):2485-96.

Garcion, E., Halilagic, A., Faissner, A. & ffrench-Constant, C. (2004) Generation of an environmental niche for neural stem cell development by the extracellular matrix molecule tenascin C. Development. 2004 Jul;131(14):3423-32.

Garwood, J., Theocharidis, U., Calco, V., Dobbertin, A. & Faissner, A. (2011) Existence of Tenascin-C Isoforms in Rat that Contain the Alternatively Spliced AD1 Domain are Developmentally Regulated During Hippocampal Development. Cell Mol Neurobiol. 2011 Oct 4. The paper has been published in (2012); add the correct designation: Cell. Mol. Neurobiol. 32(2), 279-287

Gates, M.A., Thomas, L.B., Howard, E.M., Laywell, E.D., Sajin, B., Faissner, A., Götz, B., Silver, J. & Steindler, D.A. (1995). Cell and molecular analysis of the developing and adult mouse subventricular zone of the cerebral hemispheres. J Comp Neurol, 361(2), 249-66.

Joester, A. & Faissner, A. (1999). Evidence for combinatorial variability of tenascin-C isoforms and developmental regulation in the mouse central nervous system. J Biol Chem, 274(24), 17144-51.

Michele, M. & Faissner, A. (2009) Tenascin-C stimulates contactin-dependent neurite outgrowth via activation of phospholipase C. Mol Cell Neurosci. 2009 Aug;41(4):397-408. Epub 2009 Apr 24.

Orend, G. & Chiquet-Ehrismann, R. (2006). Tenascin-C induced signaling in cancer. Cancer Lett, 244(2), 143-63.

Rigato, F., Garwood, J., Calco, V., Heck, N., Faivre-Sarrailh, C. & Faissner, A. (2002) Tenascin-C promotes neurite outgrowth of embryonic hippocampal neurons through the alternatively spliced fibronectin type III BD domains via activation of the cell adhesion molecule F3/contactin. J Neurosci. 2002 Aug 1;22(15):6596-609.

von Holst, A., Egbers, U., Prochiantz, A. & Faissner, A. (2007) Neural stem/progenitor cells express 20 tenascin C isoforms that are differentially regulated by Pax6. J Biol Chem, 282, 9172-9181.

# 6

# Lack of Evidence for Contribution of eNOS, ACE and AT1R Gene Polymorphisms with Development of Ischemic Stroke in Turkish Subjects in Trakya Region

Tammam Sipahi

*Department of Biophysics, Medical Faculty, Trakya University, Edirne*
*Turkey*

## 1. Introduction

Nitric oxide (NO) is produced in the endothelial cells, neurons, glia, and macrophages by the nitric oxide synthase (NOS) isoenzymes. Endothelial nitric oxide synthase (eNOS) is a subgroup of this family of enzymes that catalyze the production of nitric oxide (NO) from L-arginine and oxygen, which causes vascular relaxation (1) by activates guanylate cyclase, which induces smooth muscle relaxation.

The reaction catalyzed by eNOS is:

$$\text{L-arginine} + 3/2\,\text{NADPH} + H^+ + 2\,O_2 = \text{citrulline} + \text{nitric oxide} + 3/2\,\text{NADP}^+$$

NO can also promote vasorelaxation indirectly by inhibiting the release of renin which converts angiotensinogen to angiotensin I. This is in turn cleaved to form active angiotensin II by Angiotensin-converting enzyme (ACE), the key component of the physiological control of blood pressure in human. Angiotensin II exerts its effects by binding to angiotensin II type 1, 2, 3, and 4 receptors (AT1R, AT2R, AT3R, AT4R). AT1R is the major mediator of physiological effects of angiotensin II. AT1R mediates its action by association with G proteins and followed by vasoconstriction. The activated receptor in turn couples to G proteins and thus activates phospholipases, increases the cytosolic

Lack of Evidence for Contribution of eNOS, ACE and AT1R Gene Polymorphisms with
Development of Ischemic Stroke in Turkish Subjects in Trakya Region

73

$Ca^{2+}$ concentrations, which triggers cellular responses such as stimulation of protein kinases. Activated receptor also inhibits adenylyl cyclases and activates various tyrosine kinases (2).

Ischemic stroke, caused either by thrombosis or embolism, is the most frequent disease leading to disability and/or to death (3). The genetic differentiations varying with ethnic properties may be related to the arrangement of the classic and non-classic risk factors for ischemic stroke (4).

During the last two decades, there has been an increasing interest in the study of the different polymorphisms of genes of the renin-angiotensin system (RAS) and its association with the pathogenesis of stroke disease (5, 6). The RAS gene system comprises the angiotensinogen (AGT), renin, angiotensin I, angiotensin I-converting enzyme (ACE), angiotensin II, and angiotensin II receptors (7).

The ACE is a key component of both the RAS and the kinin-kallikrein system. ACE cleaves the carboxy-terminal dipeptide of angiotensin I, releasing the physiologically active octapeptide angiotensin II (8). Angiotensin II is a potent vasoconstrictive molecule that plays a key role in modulating vascular tone. Angiotensin II exerts its effects by binding to the major mediator AT1R. Human AT1R is present predominantly in vascular cells and in both kidney and adrenal gland mediating physiological actions of angiotensin II. AT1R mediates its action by association with G proteins that activate a phosphatidylinositol-calcium second messenger system, followed by vasoconstriction, hypertrophy, or catecholamine liberation at sympathetic nerve endings (9).

Our study aimed to assess the distribution of gene polymorphisms of ACE, AT1R and eNOS gene polymorphisms in ischemic stroke patients compared to healthy controls in the subjects from Trakya region.

The ACE gene maps to chromosome 17 (17q23.3), spans 21 kb, and comprises 26 exons and 25 introns, and is characterized by a polymorphism resulting from the presence (insertion) or absence (deletion) of a 287 base pairs fragment of a repeated Alu sequence at intron 16 hence, the corresponding designation of insertion (I) or deletion (D) of the two resulting alleles (10, 11).

The AT1R gene maps to chromosome 3 (3q21q25), spans 45.123 kb, and comprises 5 exons and 4 introns (12). AT1R entire coding region harbored only on exon 5, and is characterized by a polymorphism resulting from an A/C (adenine/cytosine) transversion located at position 1166 (A1166C polymorphism) in 3' untranslated region (13).

The eNOS gene is located on chromosome 7q35-36 and comprises 26 exons spanning 21 kb (14). Three classes of genetic polymorphisms in eNOS have been identified: those in intron regions, those in the promoter, and those in exon regions (15).

The variable number of tandem repeat (27 VNTR) polymorphism in intron 4 of the eNOS gene (eNOS 4 a/b), and Guanine (G) to Thymine (T) conversion at nucleotide position 894 in exon 7 causing Glutamic acid (Glu) to Aspartic acid (Asp) change at 298 are two of the most encountered polymorphisms. This polymorphism was shown to affect the response of vascular endothelium and the NO levels of plasma (16, 17).

In view of the aging population stroke is becoming a major problem, it is the most frequent disease leading to disability (3) and estimates forecast a continuing increase in the incidence, prevalence, and mortality of stroke in the next decades.

## 2. Material and methods

The study included 341 subjects; 197 stroke patients and 144 controls **(Table 1)**. All participants gave informed consent that was approved by the local ethics committee. DNA was isolated from peripheral blood, collected into tubes containing ethylenediamine-tetraacetic acid (EDTA) by eZNA (EaZy Nucleic Acid Isolation) blood DNA kits (Omega Bio-tek, Doraville, USA). eNOS (4 a/b) and ACE (I/D) gene polymorphisms were identified using a polymerase chain reaction (PCR) technique (5, 18). The AT1R (A1166C) and eNOS (Glu298Asp) gene polymorphisms were identified using PCR technique and restriction fragment length polymorphism (RFLP) assay (5, 19).

|  | Control Group | Stroke Group | p |
|---|---|---|---|
| Hypertension (%) | 61.3 | 83.1 | <0.001 |
| Current smoker (%) | 3.6 | 28.3 | <0.001 |
| Diabetes mellitus (%) | 17.6 | 33.8 | 0.001 |
| Family history of stroke (%) | 17.6 | 33.0 | 0.002 |
| Age (years) | 63.0 (17.0) | 69.0 (14.0) | <0.001 |
| SBP (mmHg) | 120.0 (20.0) | 140.0 (40.0) | <0.001 |
| DBP (mmHg) | 70.0 (10.0) | 80.0 (20.0) | <0.001 |
| FBG (mg/dl) | 89.5 (18.3) | 105.5 (41.0) | <0.001 |
| TG (mg/dl) | 117.5 (93.5) | 145.0 (105.0) | 0.008 |
| TC (mg/dl) | 189.0 (44.0) | 190.0 (52.0) | NS |
| HDL-C (mg/dl) | 39.0 (19.5) | 38.5 (14.0) | NS |
| LDL-C (mg/dl) | 120.5 (35.0) | 124.0 (41.5) | NS |

SBP/DBP; Systolic/Diastolic blood pressure, FBG; Fasting blood glucose, TG; Triglycerides, TC; Total cholesterol, HDL-C/LDL-C; High/Low density lipoprotein cholesterol, NS: Non-significant.

Table 1. Demographic and clinical characteristics of the control and stroke groups

PCR technique, developed in 1983 by Kary Mullis, is an in vitro indispensable scientific technique used in medical genetics and hereditary disorders researches to amplify a single (or a few copies) of a piece of DNA to generating millions of copies of a particular DNA sequence (20).

The method relies on thermal cycling, consisting of steps of thermal cycling which can be accomplished automatically with the DNA thermal cycler. First step is DNA denaturation. DNA denaturation is necessary first to physically separate the two strands in a DNA double helix at a high temperature in a process called DNA melting. The four bases found in DNA are adenine (A), cytosine (C), guanine (G) and thymine (T). A base on one strand normally binds only to T on the other strand, and C base on one strand normally binds only to G on the other strand. The two types of base pairs form different numbers of hydrogen bonds, AT forming two hydrogen bonds, and GC forming three hydrogen bonds. To separate the tow strands of DNA, typical strand separation temperatures ($T_{ss}$) are 95°C for 30 seconds, or 97°C for 15 seconds (21). For G and C rich region higher temperature may be appropriate (21). The second step is primer annealing. Primers contain sequences complementary to the target region of the DNA template. Primer annealing is required for initiation of DNA synthesis at a lower temperature. A temperature of 55°C is a starting degree for 20 base primers with equal GC/AT content (22). Annealing temperatures in the range of 55°C to 72°C generally yield the best results and occurs in a few seconds (21). The third step is primer extension. Primer extension depends upon the length of the target sequence. Extension at 72°C for fragments shorter than 500 base takes only 20 seconds, and fragments up to 1.2 kilo base 40 seconds is sufficient (23).

In the PCR a thermostable Taq DNA polymerase, an enzyme originally isolated from the bacterium *Thermus aquaticus,* are used. The half life of Taq DNA polymerase activity is larger than 2 hours at 92.5°C, 40 minutes at 95°C, and 5 minutes at 97.5°C (21). This DNA polymerase enzymatically assembles a new DNA strand from deoxynucleotide triphosphates (dNTPs), by using separated single-stranded DNA as a template and DNA primers. Because the primer extension products synthesized in one cycle can serve as a template in the next, the DNA template is exponentially amplified. Thus, 20 cycles of PCR yields about a million - fold ($2^{20}$) amplification (22). Since strand dissociation temperatures, primer annealing, product specificity, and Taq DNA polymerase activity affected by magnesium concentration, the magnesium ion concentration was optimized for all gene amplifications in the study. Also, a recommended buffer for PCR is 10 to 50 mM Tris-HCl pH 8.3, up to 50 mM KCl, and up to 0.1% detergents such as Tween 20 must be included. The PCR products of a particular segment of DNA in an ethidium bromide stained agarose gel visualized by UV transillumination. The minimum amount which can be detected by UV transillumination is larger than 10 ng DNA.

Restriction endonucleases are a set of enzymes expressed in bacteria against foreign DNA. Restriction enzymes cut or cleave double stranded DNA at specific recognition base sequences. In 1970 Smith H. et al identified the first restriction enzyme Hind II. Over 3000 of restriction enzymes have been isolated from different bacterial species (24, 25). Restriction enzymes can be used to distinguish single base changes in DNA (26). This method can be used to genotype a DNA sample without the need for expensive gene sequencing. The sample is first digested with the restriction enzyme to generate DNA fragments, and then the different sized fragments separated by gel electrophoresis. The choice of a restriction enzyme for PCR product is dictated by the product itself. All restriction enzymes require $Mg^{2+}$ ions as a cofactor and 37°C is optimal for most of them to works. The recommended units and digestion buffer for 100% digestion with restriction enzymes is 10-20 units for 0.1

to 0.5 µg of PCR products and 10 mM Tris-HCl (pH 7.5 to 8.5), 10 mM MgCl$_2$, 100 mM KCl and 0.1 mg/mL BSA.

A genomic DNA were amplified by PCR technique in a total 25 µL PCR mixture containing 200 ng of DNA, deoxynucleotide triphosphates (0.2 mM of each), 0.5 nmol of sense and anti-sense oligonucleotide primers, 1X Taq buffer and 1.25 U of Taq DNA polymerase. eNOS (4a/b), ACE (I/D) and eNOS (Glu298Asp) gene polymorphism reactions were contained 2.5 mM MgCl$_2$ whereas AT1R (A1166C) gene polymorphism reaction were contained 1.5 mM MgCl$_2$. All reagents for PCR amplification and gel electrophoresis were purchased from Fermentas Life Sciences (ELİPS, Istanbul, Turkey). All other chemicals were from Sigma and Merck (BO&GA, Istanbul, Turkey) and of the highest purity available. DNA amplifications were performed with a Techne (TechGene) DNA Thermal Cycler.

## 3. ACE I/D gene polymorphism (rs 4646994)

The PCR primers with the sequences reported by Rigat B. et al. (27) were used. Sense and anti-sense primers were; 5′-CTGGAGACCACTCCCATCCTTTCT-3′ and 5′-GATGTGGCCATCACATTCGTCAGAT-3′, respectively. Normally the sense primer in Rigat et al. didn't contain (G), so our PCR products also didn't contain (G), and the anti-sense primer in Rigat et al. didn't contain (G); which is normally must be included, but it was instead of (G) contained (A). Figure 1 shows the sequencing of the region which contains ACE (D) polymorphism.

**PCR Conditions**

| | | |
|---|---|---|
| 94°C | 5 min | |
| **94°C** | **1 min** | |
| **58°C** | **1 min** | **30 Cycles** |
| **72°C** | **1 min** | |
| 72°C | 7 min | |

The expected insertion (I) and deletion (D) alleles were visualized after electrophoresis on a 2% agarose gel and ethidium bromide staining under UV light transillumination (**Fig. 2**). Preferential amplification of the D allele in the heterozygotes has led to their mistyping as DD homozygotes (28). To exclude this possibility, all DD homozygotes were retyped using I

*CTGGAGA(G)CCACTCCCATCCTTTCT*CCCATTTCTCTAGACCTGCTGCCTATACAG

TCACTTTTATGTGGTTTCGCCAATTTTATTCCAGCTCTGAAATTCTCTGAGCTCCCC

TTACAAGCAGAGGTGAGCTAAGGGCTGGAGCTCAAGGCATTCAAACCCCTACCA

*GATCTGACGAATGTGATGGCCAC(G→A)TC*

Fig. 1. The sequencing of the region which contains ACE (D) polymorphism. Italic and bold letters were used for the primer sequences.

allele specific sense primer 5'-TTTGAGACGGAGTCTCGCTC-3' and anti-sense primer, also
reported by Rigat B et al. (27) were used. Amplification was performed with a DNA
Thermal Cycler with 3 min of denaturation at 93°C, followed by 30 cycles with 1 min of
denaturation at 93°C, annealing for 1.5 min at 68°C, and extension for 2 min at 72°C,
followed by 3 min of extension at 72°C. When a DD sample amplified using the I-specific
primer, it was retyped ID.

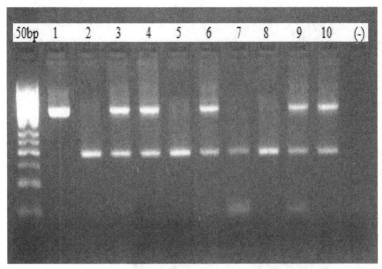

Fig. 2. PCR products of ACE gene I/D polymorphism. The DD (190 bp; lane 2, 5, 7, and 8),
the ID (190 bp, and 490 bp, lane 3, 4, 6, 9, and 10) and the II (490 bp, lane 1), 50 bp is a size
marker, (-) is a negatif control.

## 4. eNOS 4 a/b (27 VNTRs) gene polymorphism

The PCR primers with the sequences reported by Wang et al. (29) were used. Sense and anti-
sense         primers         were;         5'-AGGCCCTATGGTAGTGCCTT-3'         and
5'-TCTCTTAGTGCTGTGGTCAC-3', respectively. Figure 3 shows the sequencing of the
region which contains eNOS 4 a/b (27 VNTRs) polymorphism.

**PCR Conditions**

| | |
|---|---|
| 94°C | 1 min |
| **95°C** | **25 sec** ⎤ |
| **56°C** | **35 sec** ⎬ **38 Cycles** |
| **72°C** | **40 sec** ⎦ |
| 72°C | 5 min |

The PCR products were electrophorized on 2.5% agarose gels, stained with ethidium
bromide, and checked under UV light transillumination **(Fig. 4)**.

*AGGCCCTATGGTAGTGCCTT*GGCTGGAGGAGGGGAAA**GAAGTCTAGACCTGCTG**

**CAGGGGTGAG**GAAGTCTAGACCTGCTGCAGGGGTGAGGAAGTCTAGACCTGCTG

CAGGGGTGAGGAAGTCTAGACCTGCTGCGGGGGTGAGGAAGTCTAGACCTGCTG

CGGGGGTGAGGACAGCTGAGCGGAGCTTCCCTGGGCGGTGCTGTCAGTAGCAGG

AGCAGCCTCCTGGAAAAGCCCTGGCTGCTGCTTCTCCCCCAAGAGAGAAGGCTTC

TCCCGCCAGGCCAGTCCAGTGCAGCCCCTCACCCACACCCACTGCTACCCCAGTT

CCCCTGCTTCGGCCCGCACCCTCCCTCACACCCCAGCCCACAGACTCGGGGCTGG

CCTTAGTTACTGGAACGCCT*GTGACCACAGCACTAAGAGA*

Fig. 3. The sequencing of the region which contains eNOS 4 a/b (27 VNTRs) polymorphism. Italic letters were used for the primer sequences and bold letters were used for 27-bp repeats which are deleted in the VNTR 4a polymorphism.

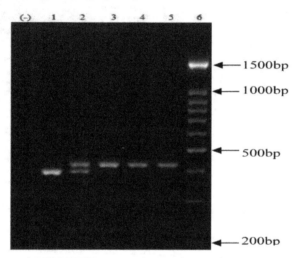

Fig. 4. PCR products of eNOS VNTR gene polymorphism. The aa genotype (394 bp; lane 1), the ab genotype (394 bp and 421 bp; lane 2), and the bb genotype (421 bp; samples 3, 4, and 5). Lane (-) is a negative control, and 6 is a size marker (O'RangeRuler 100bp DNA Ladder).

## 5. AT1R A1166C gene polymorphism (rs 5186)

AT1R A1166C gene polymorphism was identified with PCR technique followed by RFLP with the restriction enzyme HaeIII (30).

PCR primers were generated to amplify the 255 bp fragment encompassing the A1166C variant (sense and anti-sense primers were 5′-GCAGCACTTCACTACCAAATG$\underline{G}$GC-3′ and 5′-CAGGACAAAAGCAGGCTAGGGAGA -3′, respectively) in a 25 µL PCR mixture. Figure 5 shows the sequencing of the region which contains AT1R A1166C gene polymorphism. The sense primer contains one mismatch (A→G) which was required for restriction site.

Lack of Evidence for Contribution of eNOS, ACE and AT1R Gene Polymorphisms with
Development of Ischemic Stroke in Turkish Subjects in Trakya Region

79

## PCR Conditions

| 94°C | 5 min |
|------|-------|

| 94°C | 1 min |
|------|-------|
| 55°C | 1 min |
| 72°C | 1 min |

35 Cycles

| 72°C | 7 min |
|------|-------|

The PCR products were electrophorized on 2% agarose gels, stained with ethidium bromide, and checked under UV light transillumination.

*GCAGCACTTCACTACCAAATG(A→G)GCA*TTAGCTACTTTTCAGAATTGAAGGAGA

AAATGCATTATGTGGACTGAACCGACTTTTCTAAAGCTCTGAACAAAAGCTTTTC

TTTCCTTTTGCAACAAGACAAAGCAAAGCCACATTTTGCATTAGACAGATGACGG

CTGCTCGAAGAACAATGTCAGAAACTCGATGAATGTGTTGATTTGAGAAATTTTA

CTGACAGAAATGCAA*TCTCCCTAGCCTGCTTTTGTCCTG*

Fig. 5. The sequencing of the region which contains AT1R A1166C polymorphism. Italic and bold letters were used for the primer sequences. The underlined and bold letters represent the restriction site for HaeIII (5′-GG↓CC-3′).

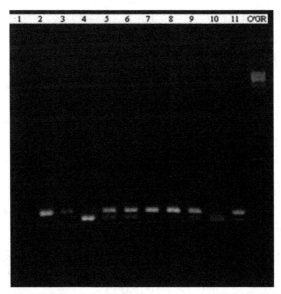

Fig. 6. EtBr stained gel of HaeIII digested PCR products of AT1R A1166C shows the AA genotype (255 bp; lane 2, 3, 7, and 8), the AC genotype (255 bp, 231 bp, and 24 bp; lane 5, 6, 9, and 11), the CC genotype (231 bp, and 24bp; lane 1, 4, and 10), lane O'GR is a size marker (100bp DNA Ladder).

Ten microliters of PCR product were digested with 5 unite of the restriction enzyme HaeIII (Takara Bio Inc, Japan) in 1 X M buffer (10 mM Tris-HCl, pH 7.5, 10 mM MgCl$_2$, 1 mM Dithiothreitol and 50 mM NaCl) for 2 hours at 37°C. When mutant allele (cytosine), digested with HaeIII that yield two fragments, whereas a wild allele (adenine) at nucleotide position 1166, had no cutting site for HaeIII, so that the PCR product was not cleaved into two fragments. The restriction digest products were visualized after electrophoresis on a 2.5% agarose gel and ethidium bromide staining (Fig. 6).

## 6. eNOS Glu298Asp (rs 1799983) gene polymorphism

Glu298Asp polymorphism of eNOS was identified with PCR technique followed by RFLP with the restriction enzyme BanII (19, 31).

PCR primers were generated to amplify the 248 bp fragment encompassing the eNOS Glu298Asp variant primers 5'-AAGGCAGGAGACAGTGGATGGA-3' (sense) and 5'-CCC AGTCAATCCCTTTGGTGCTCA-3' (anti-sense). Figure 7 shows the sequencing of the region which contains eNOS Glu298Asp gene polymorphism.

**PCR conditions**

### eNOS Glu298Asp;

| | | |
|---|---|---|
| 95°C | 5 min | |
| **94°C** | **1 min** | ⎫ |
| **59°C** | **1 min** | ⎬ 38 Cycles |
| **72°C** | **1 min** | ⎭ |
| 72°C | 5 min | |

The PCR products were electrophorized on 2% agarose gels, stained with ethidium bromide, and checked under UV light transillumination.

*AAGGCAGGAGACAGTGGATGGA*GGGGTCCCTGAGGAGGGCATGAGGCTCAGCCC

CAGAACCCCCTCTGGCCCACTCCCCACAGCTCTGCATTCAGCACGGCTGGACCCC

AGGAAACGGTCGCTTCGACGTGCTGCCCCTGCTGCTGCAGGCCCCAGAT**GATCCC**

**C**CAGAACTCTTCCTTCTGCCCCCCGAGCTGGTCCTTGAGGTGCCCCTGGAGCACCC

CACG*TGAGCACCAAAGGGATTGACTGGG*

Fig. 7. The sequencing of the region which contains eNOS Glu298Asp polymorphism. Italic and bold letters were used for the primer sequences. The underlined and bold letters represent the restriction site for Ban II (5'-G(A/G)GC(T/C)↓C-3').

Ten microliters of PCR product were digested with the restriction enzyme BanII to digest wild allele (guanine). When a guanine is at nucleotide position 894, resulting in a glutamic acid at amino acid position 298, BanII restriction enzyme produces two fragments of 163 and

Lack of Evidence for Contribution of eNOS, ACE and AT1R Gene Polymorphisms with
Development of Ischemic Stroke in Turkish Subjects in Trakya Region

81

85 bp. In contrast, when a thymine is at nucleotide position 894 (mutant allele), resulting in an aspartic acid in the amino acid sequence, the Asp298 variant had no cutting site for BanII, so that the 248 bp PCR product was not cleaved into 163 and 85 bp fragments. The restriction digest products were analyzed through electrophoresis on 2.5% agarose gel and ethidium bromide staining (Fig. 8).

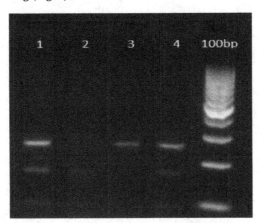

Fig. 8. EtBr stained gel of BanII digested products of eNOS gene Glu298Asp polymorphism. Line 1 and 4; GT alleles (85, 163, and 248 bp), line 2; GG alleles (85 and 163 bp), line 3; TT alleles (248 bp) and line 100bp; Gene Ruler 100 bp DNA Ladder.

## 7. Results and discussion

Table 2, 3, 4, and 5 shows the distributions of ACE I/D, eNOS (4 a/b), AT1R (A1166C), and eNOS Glu298Asp genotypes, respectively.

Statistical analyses were performed with the SPSS 15.0 software and STATA program. Genotypic distributions were in accordance with Hardy-Weinberg equilibrium in the stroke group as well as in the control group. Several studies have shown differences in the genotypic distributions of these genes while, others have shown no differences between the controls and patients. Our results didn't show any significant difference between the ischemic stroke patients and the controls ($p > 0.05$) and suggested the lack of an association between the 4 gene polymorphisms and ischemic stroke (Table 2, 3, 4, and 5). So the 4 gene polymorphisms did not enhance the predictability of stroke.

|                     | DD              | ID              | II              |
|---------------------|-----------------|-----------------|-----------------|
| Controls (%)        | 34.3            | 49.7            | 16.1            |
| Stroke Patients (%) | 34.0            | 50.0            | 16.0            |
|                     | Non-Significant | Non-Significant | Non-Significant |

Table 2. Distribution of ACE (I/D) genotype frequency in the controls and stroke patients

|                      | aa              | ab              | bb              |
|----------------------|-----------------|-----------------|-----------------|
| Controls (%)         | 2.8             | 29.8            | 67.4            |
| Stroke Patients (%)  | 2.0             | 35.0            | 63.0            |
|                      | Non-Significant | Non-Significant | Non-Significant |

Table 3. Distribution of eNOS (4 a/b) genotype frequency in the controls and stroke patients

|                      | AA              | AC              | CC              |
|----------------------|-----------------|-----------------|-----------------|
| Controls (%)         | 60.1            | 35.7            | 4.2             |
| Stroke Patients (%)  | 58.0            | 34.6            | 7.4             |
|                      | Non-Significant | Non-Significant | Non-Significant |

Table 4. Distribution of AT1R (A1166C) genotype frequency in the controls and stroke patients

|                      | GG              | GT              | TT              |
|----------------------|-----------------|-----------------|-----------------|
| Controls (%)         | 49.3            | 45.8            | 4.9             |
| Stroke Patients (%)  | 56.3            | 40.6            | 3.1             |
|                      | Non-Significant | Non-Significant | Non-Significant |

Table 5. Distribution of eNOS (Glu298Asp) genotype frequency in the controls and stroke patients

In our previous study about potential angiotensinogen (AGT) gene that predispose to hypertension, we failed to detect any relation between T174M and M235T gene polymorphisms of the AGT gene in the RAS and the development of hypertension (32).

Now, we are working on the AGT gene to clarify the role of T174M and M235T gene polymorphisms of the AGT gene in the stroke Turkish patients from Trakya region.

## 8. Conclusions

In addition to demographic and clinical characteristics, which are important in the developing of ischemic stroke, our data does not suggest that ACE (I/D), AT1R (A1166C), eNOS (4 a/b) and eNOS (Glu298Asp) gene polymorphisms, in contrast to other studies which shows a positive association between this gene polymorphisms and ischemic stroke, are a common cause of ischemic stroke in Turkish patients from Trakya region.

## 9. References

[1] Arnal JF, Dinh-Xuan AT, Pueyo M et al. Endothelium-derived nitric oxide and vascular physiology and pathology. Cell Mol Life Sci 55: 1078-1087. doi: 10.1007 / s000180050358, (1999).

[2] Matsusaka T, Ichikawa I. Biological functions of angiotensin and its receptors. Annu. Rev. Physiol. 59: 395–412. doi:10.1146/annurev.physiol.59.1.395. PMID 9074770, (1997).

[3] Saver J., Tamburi T. In: Neurogenetics (S.M.Pulst, Ed.), Oxford University Press, New York, 403-432, (2000).

[4] Hassan A, Markus H. S. Genetics and ischaemic stroke. Brain 123:1784-1812. doi:10.1093/brain/123.9.1784, (2000).

[5] Sipahi T., Guldiken B., Guldiken S., Ustundag S., Turgut N., Budak M., Cakina S., Ozkan H., Sener S. The Association of Gene Polymorphisms of the Angiotensin-Converting Enzyme and Angiotensin II Receptor Type 1 with Ischemic Stroke in Turkish subjects of Trakya Region. Trakya Univ. Tip Fak. Derg. (ISI), 1-8 pp., (2009).

[6] Hong S. H., Park H. M., Ahn J. Y., Kim O. J., Hwang T. S., Oh D., et al. ACE I/D polymorphism in Korean patients with ischemic stroke and silent brain infarction. Acta Neurol Scand. 117: 244-9, (2008).

[7] de Gasparo M., Catt K. J., Inagami T., Wright J. W., Unger T. International union of pharmacology. XXIII. The angiotensin II receptors. Pharmacol Rev. 52: 415-72, (2000).

[8] Soubrier F., Hubert C., Testut P., Nadaud S., Alhenc-Gelas F., Corvol P. Molecular biology of the angiotensin I converting enzyme: I. Biochemistry and structure of the gene. J Hypertens. 11: 471-6, (1993).

[9] Kim S., Iwao H. Molecular and cellular mechanisms of angiotensin II-mediated cardiovascular and renal diseases. Pharmacol Rev. 52: 11-34, (2000).

[10] Hubert C., Houot A. M., Corvol P., Soubrier F. Structure of the angiotensin I-converting enzyme gene. Two alternate promoters correspond to evolutionary steps of a duplicated gene. J Biol Chem, 266: 15377-83, (1991).

[11] Mattei M. G., Hubert C., Alhenc-Gelas F., Roeckel N., Corvol P., Soubrier F. Angiotensin-I converting enzyme gene is on chromosome 17. Cytogenet Cell Genet. 51: 1041-5 (1989).

[12] Szpirer C., Rivière M., Szpirer J., Levan G., Guo D. F., Iwai N., et al. Chromosomal assignment of human and rat hypertension candidate genes: type 1 angiotensin II receptor genes and the SA gene. J Hypertens. 11: 919-25, (1993).

[13] Bonnardeaux A., Davies E., Jeunemaitre X., Féry I., Charru A., Clauser E., et al. Angiotensin II type 1 receptor gene polymorphisms in human essential hypertension. Hypertension. 24: 63-9 (1994).

[14] Marsden P. A., Heng H. H. Q., Scherer S. W., Stewart R. J., Hall A. V., Shi X. M., Tsui L. C., Schappert K. T. The Journal of Biological chemistry, 268 (23): 17478-17488, (1993).

[15] Wang X. L., Wang J. Molecular Genetics and Metabolism, 70: 241-251, (2000).

[16] Leeson C. P., Hingorani A. D., Mullen M. J. et al. Glu298Asp endothelial nitric oxide synthase gene polymorphism interacts with environmental and dietary factors to

influence endothelial function. Circ Res 90:1153-1158. doi:10.1161/01.RES.00000205
62.07492.D4, (2002).

[17] Metzger I. F., Serto´rio J. T., Tanus-Santos J. E. Modulation of nitric oxide formation by
endothelial nitric oxide synthase gene haplotypes. Free Radic Biol Med 43: 987–992.
doi:10.1016/j. freeradbiomed.2007.06.012, (2007).

[18] Sipahi, T. Basak A. A., Ozgen Z., Aksoy A., Omurlu I. K., Palabiyik O., Cakina S., Sener
S. Lack of Evidence for Contribution of Endothelial Nitric Oxide Synthase Intron 4
VNTR Gene Polymorphisms To Development Of Ischemic Stroke In Turkish
Subjects. Biotechnol. & Biotechnol. Eq. (ISI), 1372-1377 pp., (2009).

[19] Guldiken B., Sipahi T., Guldiken S., Ustundag S., Budak M., Turgut N., Ozkan H.
Glu298Asp polymorphism of the endothelial nitric oxide synthase gene in Turkish
patients with ischemic stroke. Mol. Biol. Rep. (ISI), 539-1543 pp., DOI:
10.1007/s11033-008-9348-7, (2009).

[20] Saiki R. K., Scharf S., Faloona F., Mullis K. B., Horn G. T., Erlich H. A., Arnheim N.
Science. 37: 170-172, (1985).

[21] Innis M. A., Gelfand D. H. Sninsky J. J., White T. J. PCR Protocols A Guide to Methods
and Applications. Academic Press, (1990).

[22] Henry A. Erlich H. A. PCR Technology Principles and Applications for DNA
Amplification. Stockton Press, (1989).

[23] Rolfs A., Schuller I., Finckh U., Weber-Rolfs I. PCR: Clinical Diagnostics and Research.
Springer-Verlag Berlin Heidelberg, (1992).

[24] Lorne T. Kirby. DNA Fingerprinting: An Introduction. W.H. Freeman and Company,
New York, (1992).

[25] Roberts R. J., Vincze T., Posfai J., Macelis D. Rebase-enzymes and genes for DNA
restriction and modification. Nucleic Acids Res 35 (Database issue): D269–70.
doi:10.1093/nar/gkl891.PMC 1899104.PMID 17202163.http://www.pubmedcentral
.nih.gov/articlerender.fcgi?tool=pmcentrez&artid=1899104, (2007).

[26] Zhang R., Zhu Z., Zhu H., Nguyen T., Yao F., Xia K., Liang D., Liu C. SNP Cutter: a
comprehensive tool for SNP PCR–RFLP assay design. Nucleic Acids Res. 33 (Web
Server issue): W489–92. doi:10.1093/nar/gki358. PMC 1160119. PMID 15980518.
http://www.pubmedcentral.nih.gov/articlerender.fcgi?tool=pmcentrez&artid=116
0119(2005).

[27] Rigat B., Hubert C., Corvol P., Soubrier F. PCR detection of the insertion/deletion
polymorphism of the human angiotensin converting enzyme gene (DCP1)
(dipeptidyl carboxypeptidase 1). Nucleic Acids Res. 20: 1433, (1992).

[28] Shanmugam V., Sell K. W., Saha B. K. Mistyping ACE heterozygotes. PCR Methods
Appl. 3: 120-1 (1993).

[29] Wang X. L., Sim A. S., Badenhap R. F., McCredle R. M., Wilcken D. E. A smoking-
dependent risk of coronary artery disease associated with a polymorphism of the
endothelial nitric oxide synthase gene. Nat. Med. 2: 41-45, (1996).

[30] Berge K. E., Berg K. Polymorphisms at the angiotensinogen (AGT) and angiotensin II
type 1 receptor (AT1R) loci and normal blood pressure. Clin Genet. 53: 214-9,
(1998).

[31] Shimasaki Y., Hirofumi Y., Michihiro Y., Masafumi N., Kiyotaka K., Hisao O., Eisaku
H., Takenobu M., Wasaku K., Yoshihiko S., Yoshihiro M., Yoshihiro O., Kazuwa N.

Association of the Missense Glu298Asp Variant of the Endothelial Nitric Oxide
Synthase Gene With Myocardial Infarction. JACC. 31: 1506-10, (1998).

[32] Basak, A. A., Sipahi T., Ustundag S., Ozgen Z., Budak M., Sen S., Sener S. Association of
Angiotensinogen T174M and M235T Gene Variants with Development of
Hypertension in Turkish Subjects of Trakya Region. Biotechnol. & Biotechnol. Eq.,
22, 984-989, (2008).

# 7

# Overview of Real-Time PCR Principles

Morteza Seifi[1,*], Asghar Ghasemi[1], Siamak Heidarzadeh[2],
Mahmood Khosravi[3], Atefeh Namipashaki[4], Vahid Mehri Soofiany[5],
Ali Alizadeh Khosroshahi[6] and Nasim Danaei[7]

[1]*Laboratory of Genetics, Legal Medicine Organization of Tabriz, Tabriz,*
[2]*Division of Microbiology, School of Public Health,*
*Tehran University of Medical Sciences, Tehran,*
[3]*Hematology Department of Medicine Faculty, Guilan University of Medical Sciences, Rasht,*
[4]*Department of Biotechnology, School of Allied Medical Sciences,*
*Tehran University of Medical Sciences, Tehran,*
[5]*Faculty of Medicine, Shahid Behesti University of Medical Sciences, Tehran,*
[6]*Jarrah Pasha Medicine Faculty of Istanbul, Istanbul,*
[7]*Department of Health and Nutrition, Tabriz University of Medical Sciences, Tabriz*
[1,2,3,4,5,7]*Iran*
[6]*Turkey*

## 1. Introduction

Real-time PCR is based on the revolutionary method of PCR, developed by Kary Mullis in the 1980s, which allows researchers to amplify specific pieces of DNA more than a billion-fold (Saiki, Scharf et al. 1985; Mullis and Faloona 1987; Mullis 1990). PCR-based strategies have propelled molecular biology forward by enabling researchers to manipulate DNA more easily, thereby facilitating both common procedures, such as cloning, and huge endeavors such as the Human Genome Project (Olson, Hood et al. 1989; Ausubel, Brent et al. 2005). Real-time PCR represents yet another technological leap forward that has opened up new and powerful applications for researchers throughout the world. This is in part because the enormous sensitivity of PCR has been coupled to the precision afforded by "real-time" monitoring of PCR products as they are generated (Valasek and Repa 2005).

Higuchi and co-workers (Higuchi, Dollinger et al. 1992; Higuchi, Fockler et al. 1993) at Roche Molecular Systems and Chiron accomplished the first demonstration of real-time PCR. By including a common fluorescent dye called ethidium bromide (EtBr) in the PCR and running the reaction under ultraviolet light, which causes EtBr to fluoresce, they could visualize and record the accumulation of DNA with a video camera. It has been known since 1966 that EtBr increases its fluorescence upon binding of nucleic acids (Le Pecq and Paoletti 1966), but only by combining this fluorescent chemistry with PCR and real-time videography could real-time PCR be born as it was in the early 1990s. Subsequently, this

---

*Corresponding Author

technology quickly matured into a competitive market, becoming commercially widespread and scientifically influential (Valasek and Repa 2005).

Real-time PCR instrumentation was first made commercially available by Applied Biosystems in 1996, after which several other companies added new machines to the market. Presently, Applied Biosystems, BioGene, Bioneer, Bio-Rad, Cepheid, Corbett Research, Idaho Technology, MJ Research, Roche Applied Science, and Stratagene all offer instrumentation lines for real-time PCR (BioInformatics 2003).

Widespread use has also resulted in a multiplicity of names for the technology, each with a different shade of meaning. Real-time PCR simply refers to amplification of DNA (by PCR) that ismonitored while the amplification is occurring. The benefit of this real-time capability is that it allows the researcher to better determine the amount of starting DNA in the sample before the amplification by PCR. Present day real-time methods generally involve fluorogenic probes that "light up" to show the amount of DNA present at each cycle of PCR. "Kinetic PCR" refers to this process as well. "Quantitative PCR" refers to the ability to quantify the starting amount of a specific sequence of DNA. This term predates real-time PCR because it can refer to any PCR procedure, including earlier gel-based end-point assays, that attempts to quantify the starting amount of nucleic acid. Rarely, one might see the term "quantitative fluorescent PCR" to designate that the quantification was accomplished via measuring output from a fluorogenic probe, although this is redundant because all of the present chemistries for real-time PCR are fluorescent. In addition, if reverse transcriptase enzymes are used before PCR amplification in any of the above situations, then "RT-PCR" replaces "PCR" in the term. Today, the two most common terms, real-time and quantitative, are often used interchangeably or in combination, because real-time PCR is quickly becoming the method of choice to quantify nucleic acids (Valasek and Repa 2005).

The basic goal of real-time PCR is to precisely distinguish and measure specific nucleic acid sequences in a sample even if there is only a very small quantity. Real-time PCR amplifies a specific target sequence ina sample then monitors theamplification progress using fluorescent technology. During amplification, how quickly the fluorescent signal reaches a threshold level correlates with the amount of original target sequence, thereby enabling quantification. In addition, the final product can be further characterized by subjecting it to increasing temperatures to determine when the double-stranded product "melts." This melting point is a unique property dependent on product length and nucleotide composition. To accomplish these tasks, conventional PCR has been coupled to state-of-the-art fluorescent chemistries and instrumentation to becomereal-time PCR (Valasek and Repa 2005).

## 2. The chemistries of real-time PCR

Today fluorescence is exclusively used as the detection method in real-time PCR. Both sequence specific probes and non-specific labels are available as reporters. In his initial work Higuchi used the common nucleic acid stain ethidium bromide, which becomes fluorescent upon intercalating into DNA (Higuchi, Dollinger et al. 1992). Classical intercalators, however, interfere with the polymerase reaction, and asymmetric cyanine dyes such as SYBR Green I and BEBO have become more popular (Bengtsson, Karlsson et al. 2003; Zipper, Brunner et al. 2004). Asymmetric cyanines have two aromatic systems containing nitrogen, one of which is positively charged, connected by amethine bridge. These dyes have virtually no fluorescence when they are free in solution due to vibrations engaging both aromatic systems, which convert electronic excitation energy into heat that dissipates to the surrounding solvent. On the other hand the dyes become brightly fluorescent when they bind to DNA, presumably to the minor groove, and rotation around the methine bond is restricted (Nygren, Svanvik et al. 1998). In PCR the fluorescence of these dyes increases with the amount of double stranded product formed, though not strictly in proportion because the dye fluorescence depends on the dye: base binding ratio, which decreases during the course of the reaction. The dye fluorescence depends also to some degree on the DNA sequence. But a certain amount of a particular double-stranded DNA target, in the absence of significant amounts of other double-stranded DNAs, gives rise to the same fluorescence every time. Hence, the dyes are excellent for quantitative real-time PCR when samples are compared at the same level of fluorescence in absence of interfering DNA. Although minor groove binding dyes show preference for runs of AT base-pairs (Jansen, Norde´n et al. 1993), asymmetric cyanines are considered sequence non-specific reporters in real-time PCR. They give rise to fluorescence signal in the presence of any double stranded DNA including undesired primer–dimer products. Primer–dimer formation interferes with the formation of specific products because of competition of the two reactions for reagents and may lead to erroneous readouts. It is therefore good practice to control for primer–dimer formation. This can be done by melting curve analysis after completing the PCR. The temperature is then gradually increased and the fluorescence is measured as function of temperature. The fluorescence decreases gradually with increasing temperature because of increased thermal motion which allows for more internal rotation in the bound dye (Nygren, Svanvik et al. 1998). However, when the temperature is reached at which the double stranded DNA strand separates the dye comes off and the fluorescence drops abruptly (Ririe, Rasmussen et al. 1997). This temperature, referred to as the melting temperature, $T_m$, is easiest determined as the maximum of the negative first derivative of the melting curve. Since primer–dimer products typically are shorter than the targeted product, they melt at a lower temperature and their presence is easily recognized by melting curve analysis (Kubista, Andrade et al. 2006).

Labeled primers and probes are based on nucleic acids or some of their synthetic analogues such as the peptide nucleic acids (PNA) (Egholm, Buchardt et al. 1992) and the locked nucleic acids (LNA) (Costa, Ernault et al. 2004). The dye labels are of two kinds: (i) fluorophores with intrinsically strong fluorescence, such as fluorescein and rhodamine derivatives (Sjöback, Nygren et al. 1995), which through structural design are brought into contact with a quencher molecule, and (ii) fluorophores that change their fluorescence properties upon binding nucleic acids. Examples of probes with two dyes are the hydrolysis probes, popularly called Taqman probes (Holland, Abramson et al. 1991), which can be based either on regular oligonucleotides or on LNA (Braasch and Corey 2001), Molecular Beacons (Tyagi and Kramer 1996; Tyagi, Bratu et al. 1998), Hybridization probes (Caplin, Rasmussen et al. 1999), and the Lion probes (http://www.biotools.net). The dyes form a donor–acceptor pair, where the donor dye is excited and transfers its energy to the acceptor molecule if it is in proximity. Originally the acceptor molecule was also a dye, but today quencher molecules are more popular (Wilson and Johansson 2003). Energy transfer and quenching are distance dependent and structural rearrangement of the probe, or, in the case of hydrolysis probes, degradation, change the distance between the donor and acceptor and, hence, the fluorescence of the system (Kubista, Andrade et al. 2006).

Probes based on a single dye, whose fluorescence changes upon binding target DNA include the LightUp probes (Svanvik, Westman et al. 2000), AllGlo probes (http://www.allelogic.com), Displacement probes (Li, Qingge et al. 2002), and the Simple probes (http://www.idahotech.com/itbiochem/simpleprobes.html).

Chemical modifications and alterations of the oligonucleotide backbone are employed in some probes to improve the binding properties to the target template. This makes it possible to use shorter probes, which is advantageous for the detection of targets with short conserved regions such as retroviruses. LightUp probes have a neutral peptide nucleic acid (PNA) backbone that binds to DNA with greater affinity than normal oligonucleotides (Kubista, Andrade et al. 2006).

The LightUp probes are 10–12 bases, which is short compared to normal oligonucleotide probes that are usually at least 25 bases (http://www.lightup.se). LNA-probes make use of modified nucleotides to enhance binding affinity. MGBprobes are hydrolysis probes with a minor groove binding molecule attached to the end of the probe to increase affinity for DNA, which makes it possible to use shorter probes (Kutyavin, Afonina et al. 2000). Examples of modified primers include: Scorpion primers (Whitcombe, Theaker et al. 1999), LUX primers (Nazarenko, Lowe et al. 2002), Ampliflour primers (Uehara, Nardone et al. 1999), and the QZyme system (BD QZymeTM Assays for Quantitative PCR, 2003). As long as a single target is detected per sample there is not much of a difference in using a dye or a probe. Assay specificity is in both cases determined by the primers. Probes do not detect primer–dimer products, but using non-optimized probe assays is hiding the problem under the rug. If primer–dimers form they cause problems whether they are seen or not. In probe based assays, particularly when high CT values are obtained, one should verify the absence of competing primer–dimer products (Kubista, Andrade et al. 2006). The traditional way is by gel electrophoresis. Recently, an alternative approach was proposed based on the BOXTO dye. BOXTO is a sequence non-specific doublestranded DNA binding dye that has distinct spectral characteristics to fluorescein and can be used in combination with FAM based probes. The BOXTO and the probe signals are detected in different channels of the real-time PCR instrument. While the probe reflects formation of the targeted product as usual, the BOXTO dye also reports the presence of any competing primer–dimer products, which can be identified by melting curve analysis (Lind, Stahlberg et al. 2006). The great advantage of probes is for multiplexing, where several products are amplified in the same tube and detected in parallel (Wittwer, Herrmann et al. 2001). Today multiplexing is mainly used to relate expression of reporter genes to that of an exogenous control gene in diagnostic applications (Mackya 2004), and for single nucleotide polymorphism (SNP) and mutation detection studies (Mhlanga and Malmberg 2001; Mattarucchi, Marsoni et al. 2005). Multiplex assays are more difficult to design because when products accumulate the parallel PCR reactions compete for reagents. To minimize competition limiting amounts of primers must be used. Also, primer design is harder, because complementarity must be avoided between all the primers. Multiplex assays can be based either on probes or on labeled primers, where labeled primers usually give rise to signal from primer–dimer products, while probes do not. The different probing technologies have their advantages and limitations. Dyes are cheaper than probes but they do not distinguish between products. Hairpin forming probes have the highest specificity, because the formation of the hairpin competes with the binding to mismatched targets. This makes them most suitable for SNP and multi-site variation (MSV) analysis (Bonnet, Tyagi et al. 1999). Hydrolysis probes require two-step PCR to function properly, which is not optimal for the polymerase reaction, and short amplicons are necessary to obtain reasonable amplification efficiencies. But they are easier to design than hairpin forming probes and an 80% success rate was recently reported (Kubista 2004).

In summary, a 'good'probe, independent of chemistry, should have low background fluorescence, high fluorescence upon target formation (high signal to noise ratio), and high target specificity. The dyes'excitation and emission spectra are important parameters to consider when designing multiplex reactions. Spectral overlap in excitation and emission should be minimized to keep cross-talk to a minimum (Kubista, Andrade et al. 2006).

## 2.1 SYBR green I

SYBR green I binds to the minor groove of dsDNA, emitting 1,000-fold greater fluorescence than when it is free in solution (Wittwer, Herrmann et al. 1997). Therefore, the greater the amount of dsDNA present in the reaction tube, the greater the amount of DNA binding and fluorescent signal from SYBR green I. Thus any amplification of DNA in the reaction tube is measured (Valasek and Repa 2005).

## 2.2 BEBO

The minor groove binding asymmetric cyanine dye BEBO is tested as sequence nonspecific label in real-time PCR. The Fluorescence intensity of BEBO increases upon binding to double-stranded DNA allowing emission to be measured at the end of the elongation phase in the PCR cycle. BEBO concentrations between 0.1 and 0.4 mM generated sufficient Fluorescence signal without inhibiting the PCR. A comparison with the commonly used reporter dye SYBR Green I shows that the two dyes behave similarly in all important aspects. The dye has absorbance and emission wavelengths that can be detected on the FAM channel on most common real-time PCR platforms, and shows a strong fluorescence increase when bound to dsDNA. BEBO can be used as an unspecific dye for real-time PCR applications or other applications where staining of dsDNA is wanted (Bengtsson, Karlsson et al. 2003).

## 2.3 BOXTO

The unsymmetrical cyanine dyes BOXTO and its positive divalent derivative BOXTO-PRO were studied as real-time PCR reporting fluorescent dyes and compared to SYBR GREEN I (SG). Unmodified BOXTO showed no inhibitory effects on real-time PCR, while BOXTO-PRO showed complete inhibition, sufficient fluorescent signal was acquired when 0.5–1.0 μM BOXTO was used with RotorGene and iCycler platforms. Statistical analysis showed that there is no significant difference between the efficiency and dynamic range of BOXTO and SG (Ahmad 2007).

## 2.4 5' nuclease (TaqMan) probes

Hydrolysis probes (also called 5'-nuclease probes because the 5'-exonuclease activity of DNA polymerase cleaves the probe) offer an alternative approach to the problem of specificity. These are likely the most widely used fluorogenic probe format (Mackay 2004) and are exemplified by TaqMan probes. In terms of structure, hydrolysis probes are sequence- specific dually fluorophore-labeled DNA oligonucleotides (Valasek and Repa 2005). One fluorophore is termed the quencher and the other is the reporter. When the quencher and reporter are in close proximity, that is, they are both attached to the same short oligonucleotide; the quencher absorbs the signal from the reporter (Valasek and Repa 2005). This is an example of fluorescence resonance energy transfer (also called Forster transfer) in which energy is transferred from a "donor" (the reporter) to an "acceptor" (the quencher) fluorophore. During amplification, the oligonucleotide is broken apart by the action of DNA polymerase (5/-nuclease activity) and the reporter and quencher separate, allowing the reporter's energy and fluorescent signal to be liberated. Thus destruction or hydrolysis of the oligonucleotide results in an increase of reporter signal and corresponds with the specific amplification of DNA (Valasek and Repa 2005). Examples of common quencher fluorophores include TAMRA, DABCYL, and BHQ, whereas reporters are more numerous (e.g., FAM, VIC, NED, etc). Hydrolysis probes afford similar precision as SYBR green I (Wilhelm, Pingoud et al. 2003), but they give greater insurance regarding the specificity because only sequence-specific amplification is measured. In addition, hydrolysis probes allow for simple identification of point mutations within the amplicon using melting curve analysis (Valasek and Repa 2005).

### 2.5 Molecular beacons

Molecular beacons are similar to TaqMan probes but are not designed to be cleaved by the 5' nuclease activity of Taq polymerase. These probes have a fluorescent dye on the 5' end and a quencher dye on the 3' end of the oligonucleotide probe. A region at each end of the molecular beacon probe is designed to be complementary to itself, so at low temperatures, the ends anneal, creating a hairpin structure. This integral annealing property positions the two dyes in close proximity, quenching the fluorescence from the reporter dye (Espy, Uhl et al. 2006). The central region of the probe is designed to be complementary to a region of the PCR amplification product. At high temperatures, both the PCR amplification product and probe are single stranded. As the temperature of the PCR is lowered, the central region of the molecular beacon probe binds to the PCR product and forces the separation of the fluorescent reporter dye from the quenching dye. The effects of the quencher dye are obviated and a light signal from the reporter dye can be detected. If no PCR amplification product is available for binding, the probe reanneals to itself, forcing the reporter dye and quencher dye together, preventing fluorescent signal (Espy, Uhl et al. 2006). Typically, a single molecular beacon is used for detection of a PCR amplification product and multiple beacon probes with different reporter dyes are used for single nucleotide polymorphism detection. By selection of appropriate PCR temperatures and/or extension of the probe length, molecular beacons will bind to the target PCR product when an unknown nucleotide polymorphism is present but at a slight cost of reduced specificity. There is not a specific temperature thermocycling requirement for molecular beacons, so temperature optimization of the PCR is simplified (Espy, Uhl et al. 2006).

### 2.6 FRET hybridization probes

FRET hybridization probes, also referred to as LightCyclerprobes; represent a third type of probe detection format commonly used with real-time PCR testing platforms. FRET hybridization probes are two DNA probes designed to anneal next to each other in a head-to-tail configuration on the PCR product. The upstream probe has a fluorescent dye on the 3' end and the downstream probe has an acceptor dye on the 5' end. If both probes anneal to the target PCR product, fluorescence from the 3' dye is absorbed by the adjacent acceptor dye on the 5' end of the second probe. The second dye is excited and emits light at a third wavelength and this third wavelength is detected. If the two dyes do not align together because there is no specific DNA for them to bind, then FRET does not occur between the two dyes because the distances between the dyes are too great. A design detail of FRET hybridization probes is the 3' end of the second (downstream) probe is phosphorylated to prevent it from being used as a primer by Taq during PCR amplification. The two probes encompass a region of 40 to 50 DNA base pairs, providing exquisite specificity (Espy, Uhl et al. 2006). FRET hybridization probe technology permits melting curve analysis of the amplification product. If the temperature is slowly raised, eventually the probes will no longer be able to anneal to the target PCR product and the FRET signal will be lost. The temperature at which half the FRET signal is lost is referred to as the melting temperature of the probe system (Espy, Uhl et al. 2006). The Tm depends on the guanine plus cytosine content and oligonucleotide length. In contrast to TaqMan probes, a single nucleotide polymorphism in the target DNA under a hybridization FRET probe will still generate a signal, but the melting curve will display a lower Tm. The lowered Tm can be characteristic for a specific polymorphism underneath the probes; however, a lowered Tm can also be the result of any sequence difference under the probes. The target PCR product is detected and the altered Tm informs the user there is a difference in the sequence being detected. Generally, more than three base pair differences under a FRET hybridization probe prevent hybridization at typical annealing temperatures and are not detected (Espy, Uhl et al. 2006). This trait of FRET hybridization probes is advantageous in cases where the genome of the organism is known to mutate at a high frequency, such as with viruses. When a single or limited number (<3) of known polymorphisms occur between two similar targets, FRET hybridization probes can also be used for discriminating strains of organisms (Espy, Uhl et al. 2006). Like molecular beacons, there is not a specific thermocycling temperature requirement for FRET hybridization probes. Molecular beacons and FRET hybridization probes, unlike TaqMan probes, are both recycled (conserved) in each round of PCR temperature cycle. Also, for Molecular beacons and FRET hybridization probes, unlike TaqMan probes, fluorescent signal does not accumulate as PCR product accumulates after each PCR cycle (Espy, Uhl et al. 2006).

## 2.7 Scorpions

Scorpions combine the detection probe with the upstream PCR primer (Whitcombe, Theaker et al. 1999) and consist of a fluorophore on the 5' end, followed by a complementary stem-loop structure (also containing the specific probe sequence), quencher dye, DNA polymerase blocker (a nonamplifiable monomer that prevents DNA polymerase extension), and finally a PCR primer on the 3' end. The probe sequence contained within the hairpin allows the scorpion to anneal to the template strand, which separates the quencher for the fluorophore and results in increased fluorescence. Because sequence-specific priming and probing is a unimolecular event, scorpions perform better than bimolecular methods under conditions of rapid cycling such as the LightCycler (Thelwell, Millington et al. 2000). Cycling is performed at a temperature optimal for DNA polymerase activity instead of the reduced temperature necessary for the 5' nuclease assay. Scorpions are specific enough for allele discrimination and may be multiplexed easily (Thelwell, Millington et al. 2000). The scorpion chemistry has been improved with the creation of duplex scorpions in which the reporter dye/probe and quencher fragment are on separate, complementary molecules (Solinas, Brown et al. 2001). The duplex scorpions still bind in a unimolecular event, but because the reporter and quenchers are on separate molecules, they yield greater signal intensity because the reporter and quencher can separate completely (Wong and Medrano 2005).

## 2.8 Sunrise™ primers

Created by Oncor (Gaithersburg, MD, USA), Sunrise primers are similar to scorpions in that they combine both the PCR primer and detection mechanism in the same molecule (Nazarenko, Bhatnagar et al. 1997). These probes consist of a dual-labeled (reporter and quencher fluorophores) hairpin loop on the 5' end, with the 3' end acting as the PCR primer. When unbound, the hairpin is intact, causing reporter quenching via FRET. Upon integration into the newly formed PCR product, the reporter and quencher are held far enough apart to allow reporter emission (Wong and Medrano 2005).

## 2.9 LUX™ fluorogenic primers

Light upon extension (LUX) primers (Invitrogen, Carlsbad, CA, USA) are self-quenched single-fluorophore labeled primers almost identical to Sunrise primers. However, rather than using a quencher fluorophore, the secondary structure of the 3' end reduces initial fluorescence to a minimal amount (Nazarenko, Lowe et al. 2002). Because this chemistry does not require a quencher dye, it is much less expensive than dual-labeled probes. While this system relies on only two oligonucleotides for specificity, unlike the SYBR Green I platform in which a dissociation curve is used to detect erroneous amplification, no such convenient detection exists for the LUX platform. Agarose gels must be run to ensure the presence of a single PCR product, a step that is extremely important not only for the LUX primers but also for the Sunrise primers and scorpions because PCR priming and probe binding are not independent in these chemistries (Wong and Medrano 2005).

## 2.10 Light-up probes

Light-up probes are peptide nucleic acids (PNAs) that use thiazole orange as the fluorophor. Upon hybridisation with DNA, duplex or triplex structures are formed with increased fluorescence intensity of the fluorophor. A quencher is not required. This technique is limited by unspecific fluorescence, which increases during PCR and therefore restricts the achievable sensitivity (Isacsson, Cao et al. 2000; Svanvik, Stahlberg et al. 2000;Svanvik, Westman et al. 2000). Some other formats use the increasing quench as indicator for product accumulation (Crockett and Wittwer 2001; Kurata, Kanagawa et al. 2001). In this case, the fluorescence is quenched by a guanine residue of the PCR product. These probes are comparatively inexpensive and easy to construct; however, measurement of the decrease of a signal is problematic, especially during the early exponential phase in which only very few probes are quenched (Wilhelm and Pingoud 2003).

## 2.11 Eclipse probe

qPCR assays using an Eclipse probe employ two primers and a sequence-specific oligonucleotide probe. The probe is complementary to a sequence within the amplicon and contains a fluorescent reporter at the 3' end, a quencher at the 5' end, and a minor groove binder (Bio-Rad Laboratories 2006). The unhybridized probe adopts a conformation that brings the reporter and quencher together, quenching the reporter. During the annealing step of PCR, the probe hybridizes to the target with the help of the minor groove binder. The probe thus becomes linearized, separating the reporter and quencher and allowing the reporter to fluoresce. The resulting fluorescent signal is proportional to the amount of amplified product in the sample (Bio-Rad Laboratories 2006).

## 2.12 Amplifluor primer

qPCR assays using Amplifluor chemistry employ two target-specific primers and one universal primer called the UniPrimer. The first target-specific primer contains a 5' extension sequence called the Z-sequence that is also found at the 3' end of the UniPrimer. The UniPrimer forms a hairpin structure (Bio-Rad Laboratories 2006). A fluorescent reporter and a quencher are attached at the 5' and the 3' ends of the stem structure, respectively. In the hairpin conformation, the reporter fluorescence is quenched due to its proximity to the quencher. During the first amplification cycle, the first target-specific primer (with the Z-sequence) hybridizes to the template and is extended. During the second amplification cycle, the second target-specific primer is used to synthesize a new target template that contains a sequence complementary to the Z-sequence. The product from the second amplification cycle can then serve as the template for the UniPrimer. In the third amplification cycle, the extended UniPrimer serves as a template for the next amplification cycle (Bio-Rad Laboratories 2006). In the fourth cycle, extension of the template through the hairpin region of the UniPrimer causes the UniPrimer to open up and adopt a linear configuration, which allows the reporter to fluoresce. Exponential amplification using the second target-specific primer and the UniPrimer occurs in subsequent amplification cycles. The resulting fluorescent signal is proportional to the amount of amplified product in the sample (Bio-Rad Laboratories 2006).

## 2.13 BD QZyme primer

qPCR assays using BD QZyme primers employ a target-specific zymogene primer, a target-specific reverse primer, and a universal oligonucleotide substrate. The oligonucleotide contains a fluorescent reporter on the 5' end and a quencher on the 3' end. When oligonucleotide substrate is intact, the fluorescence of the reporter is quenched by the quencher due to their proximity (Bio-Rad Laboratories 2006). The zymogene primer contains a sequence that encodes a catalytic DNA. During the first amplification cycle, the zymogene primer is extended. In the second cycle, the product of the first cycle is used as the template by the target-specific reverse primer, which is extended to create a new target sequence containing a catalytic DNA region. In the subsequent annealing step, the fluorescently labeled oligonucleotide substrate hybridizes to the catalytic DNA sequence and is cleaved. This cleavage separates the reporter from the quencher, resulting in a fluorescent signal that is proportional to the amount of amplified product in the sample (Bio-Rad Laboratories 2006).

## 3. Design and optimization of SYBR Green I reactions

A SYBR Green I assay uses a pair of PCR primers that amplifies a specific region within the target sequence of interest and includes SYBR Green 1 for detecting the amplified product. The steps for developing a SYBR Green I assay are:

1.   Primer design and amplicon design

2.   Assay validation and optimization

## A) Primer and Amplicon Design:

A successful real-time PCR reaction requires efficient and specific amplification of the product. Both primers and target sequence can affect this efficiency. Therefore, care must be taken when choosing a target sequence and designing primers. A number of free and commercially available software programs are available for this purpose. One popular web-based program for primer design is Primer3 (http://frodo.wi.mit.edu/cgi-bin/primer3/primer3_www.cgi). A commercially available program such as Beacon Designer software performs both primer design and amplicon selection (Bio-Rad Laboratories 2006).

## B) Guidelines of amplicon design:

1. Design amplicon to be 75–200 bp.Shorteramplicons are typically amplified with higher efficiency. An amplicon should be at least 75 bp to easily distinguish it from any primer-dimers that might form
2. Avoid secondary structure if possible. Use programs such as mfold http://www.bioinfo.rpi.edu/applications/mfold/) to predict whether an amplicon will form any secondary structure at annealing temperature. See Real-Time PCR: General Considerations (Bio-Rad bulletin 2593) for more details
3. Avoid templates with long (>4) repeats of single bases
4. Maintain a GC content of 50–60%

## C) Parameters of primer design:

1. Design primers with a GC content of 50–60%
2. Maintain a melting temperature ($T_m$) between 50°C and 65°C. We calculate $T_m$ values using the nearest-neighbor method with values of 50 mM for salt concentration and 300 nM for oligonucleotide concentration
3. Avoid secondary structure; adjust primer locations outside of the target sequence secondary structure if required
4. Avoid repeats of Gs or Cs longer than three bases
5. Place Gs and Cs on ends of primers
6. Check sequence of forward and reverse primers to ensure no 3' complementarity (avoid primer-dimer formation)
7. Verify specificity using tools such as the Basic Local Alignment Search Tool (http://www.ncbi.nlm.nih.gov/blast/)

## D) Assay Validation and Optimization:

Components a SYBR Green I qPCR reaction:

1. PCR master mix with SYBR Green I
2. Template
3. Primers

Preformulated real-time PCR master mixes containing buffer, DNA polymerase, dNTPs, and SYBR Green I dye are available from several vendors.

Optimized SYBR Green I qPCR reactions should be sensitive and specific and should exhibit good amplification efficiency over a broad dynamic range (Bio-Rad Laboratories 2006).

Steps of to determine the performance of your SYBR Green I qPCR assay:

a.  Identify the optimal annealing temperature for your assay
b.  Construct a standard curve to evaluate assay performance

### E) Annealing Temperature Optimization:

The optimal annealing temperature can easily be assessed on qPCR instruments that have a temperature gradient feature, such as the MiniOpticon™, MyiQ™, DNA Engine Opticon®, Opticon™ 2, iCycleriQ®, Chromo4™, and iQ™5 systems.

A gradient feature allows you to test a range of annealing temperatures simultaneously, so optimization reactions can be performed in a single experiment.

To find the optimal annealing temperature for reaction, recommend testing a range of annealing temperatures above and below the calculated $T_m$ of the primers.

Because SYBR Green I binds to all dsDNA, it is necessary to check the specificity of your qPCR assay by analyzing the reaction product(s). To do this, use the melt-curve function on your real-time instrument and also run products on an agarose gel. An optimized SYBR Green I qPCR reaction should have a single peak in the melt curve, corresponding to the single band on the agarose gel.

Nonspecific products that may have been co-amplified with the specific product can be identified by melt-curve analysis. In this example, the specific product is the peak with a $T_m$ of 89°C and corresponds to the upper band on the gel. The nonspecific product is the peak with a $T_m$ of 78°C and corresponds to the lower band in the gel. By comparing the gel image with the melt curve, you can identify peaks in the melt curve that correspond to specific product, additional nonspecific bands, and primer-dimers. If nonspecific products such as primer-dimers are detected by melt-curve analysis, recommend that redesign primers (Bio-Rad Laboratories 2006).

### F) Assay Performance Evaluation Using Standard Curves:

The efficiency, reproducibility, and dynamic range of a SYBR Green I assay can be determined by constructing a standard curve using serial dilutions of a known template. The efficiency of the assay should be 90–105%, the $R^2$ of the standard curve should be >0.980 or r > | –0.990 |, and the $C_T$ values of the replicates should be similar.

It is important to note that the range of template concentrations used for the standard curve must encompass the entire range of template concentration of the test samples to show that results from the test samples are within the linear dynamic range of the assay. If the test samples give results outside of the range of the standard curves, one of the following must be performed:

1.  Construct a wider standard curve covering the test sample concentrations and perform analysis to ensure that the assay is linear in that new range
2.  If the test samples give a lower $C_T$ than the highest concentration of standards used in the standard curve, repeat the assay using diluted test samples
3.  If the test samples give a higher $C_T$ than the lowest concentration of standards used in the standard curve, repeat the assay using larger amounts of the test samples

## 4. Design and optimization of TaqManProbe reactions

A TaqMan assay uses a pair of PCR primers and a dual-labeled target-specific fluorescent probe. The steps for developing a TaqMan assay are:

a.   Primer and probe design
b.   Assay validation and optimization

### A. Primer and Probe Design:

As with any qPCR reaction, TaqMan-based assays require efficient and specific amplification of the product. Typically, the primers are designed to have an annealing temperature between 55 and 60ºC. We recommend using software such as Beacon Designer for designing your TaqMan primers and TaqMan probe.. Because the dual-labeled probe is the most costly component of a TaqMan assay, suggested that order the two primers and validate their performance using SYBR Green I before ordering the dual-labeled probe.

The TaqMan probe should have a $T_m$ 5–10°C higher than that of the primers. In most cases, the probe should be <30 nucleotides and must not contain a G at its 5' end because this could quench the fluorescent signal even after hydrolysis. Choose a sequence within the target that has a GC content of 30–80%, and design the probe to anneal to the strand that has more Gs than Cs (so the probe contains more Cs than Gs).

An important aspect of designing a TaqMan probe is reporter and quencher selection. We recommend using FAM-labeled probes when designing singleplex reactions, because they are inexpensive and readily available, perform well, and can be detected by all instruments currently on the market.

Another important consideration for obtaining accurate real-time qPCR data is probe quality. Even a perfectly designed probe can fail if the probe is improperly synthesized or purified. Improper removal of uncoupled fluorescent label, inefficient coupling, and/or poor quenching can produce high fluorescent background or noise. A low signal-to-noise ratio results in decreased sensitivity and a smaller linear dynamic range. Two probes with identical sequences and identical fluorophore labels can be measurably different when synthesized by different suppliers or even at different times by the same supplier.

### B. Assay Validation and Optimization:

A TaqMan probe-based qPCR reaction contains the following components:

1.   PCR master mix
2.   Template
3.   Primers
4.   Probe(s)

Preformulated PCR master mixes containing buffer, DNA polymerase, and dNTPs are commercially available from several vendors. For TaqMan assays, we recommend using iQ™ supermix with 300 nM of each of the two primers and 200 nM of probe(s). TaqMan assays require careful attention to temperature conditions. A typical TaqMan protocol contains a denaturation step followed by a combined annealing and extension step at 55–60°C, instead of the traditional three-step PCR cycle of denaturation, annealing, and extension. This is to ensure that the probe remains bound to its target during primer extension. Typical TaqMan probes for nucleic acid quantification are designed to have a $T_m$ of 60–70°C. An optimized TaqMan assay should be sensitive and specific, and should exhibit good amplification efficiency over a broad dynamic range.

In short, construct a standard curve using dilutions of a known template and use this curve to determine the efficiency of the assay along with $R^2$ or r of the regression line. The efficiency of the reaction should be between 90 and 105%, the $R^2$ should be >0.980 or r > |– 0.990|, and the replicates should give similar $C_T$ values. If the assay performs within these specifications, you are ready to start your experiment. If the assay performs outside these specifications, we suggest that you redesign your primers and TaqMan probe. It is important to note that the range of template concentrations used for the standard curve must encompass the entire range of template concentrations of the test samples to demonstrate that results from the test samples are within the dynamic range of the assay (Bio-Rad Laboratories 2006). If test samples give results outside the range of the standard curve, one of the three following steps must be performed:

1.  Construct a wider standard curve covering the test sample concentrations and perform analysis to ensure that the assay is linear in that new range
2.  If the test samples give a lower $C_T$ than the highest concentration of standards used in the standard curve, repeat the assay using diluted test samples
3.  If the test samples give a higher $C_T$ than the lowest concentration of standards used in the standard curve, repeat the assay using larger amounts of the test samples

## 5. The instrumentation of real-time PCR

A critical requirement for real-time PCR technology is the ability to detect the fluorescent signal and record the progress of the PCR. Because fluorescent chemistries require both a specific input of energy for excitation and a detection of a particular emission wavelength, the instrumentation must be able to do both simultaneously and at the desired wavelengths. Thus the chemistries and instrumentation are intimately linked (Valasek and Repa 2005).

At present, there are three basic ways in which real-time instrumentation can supply the excitation energy for fluorophores: by lamp, light-emitting diode (LED), or laser. Lamps are classified as broad-spectrum emission devices, whereas LEDs and lasers are narrow spectrum. Instruments that utilize lamps (tungsten halogen or quartz tungsten halogen) may also include filters to restrict the emitted light to specific excitation wavelengths. Instruments using lamps include Applied Biosystem's ABI Prism 7000, Stratagene's Mx4000 and Mx3000P, and Bio-Rad's iCycleriQ. LED systems include Roche's LightCycler, Cepheid's SmartCycler, Corbett's Rotor-Gene, and MJ Research's DNA Engine Opticon 2. The ABI Prism 7900HT is the sole machine to use a laser for excitation (Valasek and Repa 2005). To collect data, the emission energies must also be detected at the appropriate wavelengths. Detectors include charge-coupled device cameras, photomultiplier tubes, or other types of photodetectors. Narrow wavelength filters or channels are generally employed to allow only the desired wavelength(s) to pass to the photodetector to be measured. Usually, multiple discrete wavelengths can be measured at once, which allows for multiplexing, i.e., running multiple assays in a single reaction tube (Valasek and Repa 2005). Another portion of the instrumentation consists of a thermocycler to carry out PCR. Of particular importance for real-time PCR is the ability of the thermocycler to maintain a consistent temperature among all sample wells, as any differences in temperature could lead to different PCR amplification efficiencies. This is accomplished by using a heating block (Peltier based or resistive), heated air, or a combination of the two. As one might expect, heating blocks generally change temperature more slowly than heated air, resulting in longer thermocycling times. For example, Roche's LightCycler models utilizing heated air can perform 40 cycles in 30 min, whereas Applied Biosystem's ABI Prism 7900HT utilizing a Peltier-based heating block take s 1 h 45 min (Valasek and Repa 2005). Real-time instrumentation certainly would not be complete without appropriate computer hardware and data-acquisition and analysis software. Software platforms try to simplify analysis of real-time PCR data by offering graphical output of assay results including amplification and dissociation (melting point) curves. The amplification curve gives data regarding the kinetics of amplification of the target sequence, whereas the dissociation curve reveals the characteristics of the final amplified product (Valasek and Repa 2005).

## 5.1 Comparison of the different systems

Essentially, each real time PCR instrument consists of a computer-controlled thermocycler integrated with fluorescent detection system and dedicated software to analyze the result. Some systems can detect four different wave lengths (I-cycler, Mx4000 [stratagene] and Smart Cycler®, Version 2.0 Light Cycler®) whereas others can detect two different wavelengths (Light Cycler®). The Light Cycler® and Smart Cycler® are capable of performing rapid-cycle real time PCR because the reaction is set-up in capillaries or especially designated tubes. Both have optimized heating- cooling characteristic. A complete amplification protocol can be performed in 30-45 minutes (Myi ; Giulietti, Overbergh et al. 2001; Soheili and Samiei 2005). The Smart Cycler® is a combination of 16 individual, one tube real time PCR units. It is capable of performing a different PCR program on each of 16 reaction tubes. This is very useful for a rapid optimization of the assay as many variables can be tested at the same time. The Bio-Rad I-cycler IQ® instrument can perform real time amplification with a temperature gradient for specific PCR steps, allowing the optimization of real time PCR assay. The spectrofluorometers in the thermal cycler have a number of differences. Laser-based systems are tuned to excite each fluorophore at a specific wavelength and provide maximum efficiency. Lamp-based systems provide a broad excitation range that can be filtered to work with a number of fluorophores. The laser source not only gives brighter illumination to the fluorophore signal, but also produces less background noise(Myi ; Giulietti, Overbergh et al. 2001; Soheili and Samiei 2005).

In conclusion, real time PCR is a powerful advancement of the basic PCR technique. The important steps in deciding which particular assay format to use are related to the type of data required. The requirement for a research laboratory is quite distinct from those of a diagnostic laboratory. For the latter, probe confirmation of the PCR product is an essential part of the assay, whereas SYBR green detection may be sufficient for many other applications such as quantifying expression of a gene. All of the real-time PCR machines analyzed are capable of detecting PCR product in real time and a specific assay can be made optionally on every system. However, there are some decisions to be made when selecting among different formats. The choice of system is dependent on individual laboratory needs (Myi ; Giulietti, Overbergh et al. 2001; Soheili and Samiei 2005). Considering diagnostic applications, the Light Cycler® or Smart Cycler® may obtain faster results for urgent assays.

This could reduce the time of analysis to result from 3-4 hours to 1.5 hours. On the other hand, if sensitivity is the most important issue, these machines, with their smaller reaction volume and consequently lower sensitivity, wouldn't be the first choice. The ABI 7700 and Bio-Rad -I-Cycler IQ® have a 96 well format, enabling higher throughput than other systems. The 384-well plates, as designed by ABI for use in the 7900 HT system, can further enhance through put. For diagnostic application, internal control of nucleic acid isolation and PCR inhibition, it is essential to obtain valid results. This can be achieved using the system that enables multi-color detection, such as the I-Cycler IQ® and the Smart Cycler®. Recently, a multi-color format of the Light Cycler® is also present in market. Multiplex real-time PCRs can be developed for three different targets and an internal control by using the four detection wavelengths possible in multicolor detection. As a matter of fact, the choice of which real time system to use depends on the range of application required. To achieve meaningful results, each assay must be validated and optimized for the particular system chosen (Myi ; Giulietti, Overbergh et al. 2001; Soheili and Samiei 2005).

## 6. Advantages of real-time PCR quantitation

There are many methods in molecular biology for measuring quantities of target nucleic acid sequences. However, most of these methods exhibit one or more of the following shortcomings: they are time consuming, labor intensive, insufficiently sensitive, non-quantitative, require the use of radioactivity, or have a substantial probability of cross contamination (Reischl, Wittwer et al. 2002). These methods include but are not limited to Northern and Southern hybridizations, HPLC, scintillation proximity assay, PCR-ELISA, RNase protection assay, in situ hybridization, and various gel electrophoresis PCR end-point systems (Valasek and Repa 2005). Real-time PCR has distinct advantages over these earlier methods for several reasons. Perhaps the most important is its ability to quantify nucleic acids over an extraordinarily wide dynamic range (at least 5 log units). This is coupled to extreme sensitivity, allowing the detection of less than five copies (perhaps only one copy in some cases) of a target sequence, making it possible to analyze small samples like clinical biopsies or miniscule lysates from laser capture microdissection. With appropriate internal standards and calculations, mean variation coefficients are 1–2%, allowing reproducible analysis of subtle gene expression changes even at low levels of expression (Klein 2002; Luu-The, Paquet et al. 2005). In addition, all real-time platforms are relatively quick, with some affording high-throughput automation. Finally, real-time PCR is performed in a closed reaction vessel that requires no post-PCR manipulations, thereby minimizing the chances for cross contamination in the laboratory (Valasek and Repa 2005).

## 7. Limitations of real-time PCR quantitation

There are several limitations to real-time PCR methods. The majority of these are present in all PCR or RT-PCR-based techniques. Real-time PCR is susceptible to PCR inhibition bycompounds present in certain biological samples. For example, clinical and forensic uses for real-timePCR may be affected by inhibitors found in certain body fluidssuch as hemoglobin or urea (Wilson 1997). Food microbiological applications may encounter organic and phenolic inhibitors (Wilson 1997). To circumvent this problem, alternative DNA polymerases (e.g., Tfl, Pwo, Tth, etc.) that are resistant to particular inhibitors can be used. Other limitations primarily concern real-time PCR- based analysis of gene expression

(Bustin 2000; Bustin 2002;Bustin and Nolan 2004). Because of the necessary use of RNA in an extra enzymatic step, more problems have the opportunity to occur. RNA itself is extremely labile compared with DNA, and therefore isolation must be carefully performed to ensure both the integrity of the RNA itself and the removal of contaminating nucleases, genomic DNA, and RT or PCR inhibitors. This can be a problem with any sample source, but clinical samples are of special concern because inconsistencies in sample size, collection, storage, and transport can lead to a variable quality of RNA templates. Conversion of RNA to cDNA during the RT reaction is also subject to variability because multiple reverse transcriptase enzymes with different characteristics exist, and different classes of oligonucleotides (e.g., random, poly-dT, or gene specific primers) can be used to prime RT (Valasek and Repa 2005). Probably the largest present limitation of real-time PCR, however, is not inherent in the technology but rather resides in human error: improper assay development, incorrect data analysis, or unwarranted conclusions. In our experience using real-time PCR for gene expression analysis, real-time PCR primer sets must be designed and validated by stringent criteria to ensure specificity and accuracy of the results. For microbiology, false positives or negatives must be considered when designing an assay to detect pathogens. Amplification and melting curves must be visually inspected while independent calculations based on these curves should be double-checked for accuracy. Real-time PCR gene expression analysis measures mRNA levels and, therefore, only suggests possible changes in protein levels or function rather than demonstrating them. And although there is a tight connection between gene expression and gene product function (Brown and Botstein 1999)(8), this is certainly not always the case, and formal demonstration may be needed for a given research project. Of course, conclusions based on data derived from real-time PCR are best utilized when the biological context is well understood (Bustin 2002).

## 8. Types of real-time quantification

### 1. Absolute Quantitation

Absolute quantitation uses serially diluted standards of known concentrations to generate a standard curve. The standard curve produces a linear relationship between Ct and initial amounts of total RNA or cDNA, allowing the determination of the concentration of unknowns based on their Ct values (Heid, Stevens et al. 1996). This method assumes all standards and samples have approximately equal amplification efficiencies (Souaze, Ntodou-Thome et al. 1996). In addition, the concentration of serial dilutions should encompass the levels in the experimental samples and stay within the range of accurately quantifiable and detectable levels specific for both the real-time PCR machine and assay.The PCR standard is a fragment of double-stranded DNA (dsDNA), single-stranded DNA (ssDNA), or cRNA bearing the target sequence (Wong and Medrano 2005). A simple protocol for constructing a cRNA standard for one-step PCR can be found in Fronhoffs et al. (Fronhoffs, Totzke et al. 2002), while a DNA standard for two-step real-time PCR can be synthesized by cloning the target sequence into a plasmid (Gerard, Olsson et al. 1998), purifying a conventional PCR product (Liss 2002), or directly synthesizing the target nucleic acid. The standard used must be a pure species. DNA standards have been shown to have a larger quantification range and greater sensitivity, reproducibility, and stability than RNA standards (Pfaffl, Tichopad et al. 2004). However, a DNA standard cannot be used for a one-step real-time RT-PCR due to the absence of a control for the reverse transcription efficiency (Giulietti, Overbergh et al. 2001).

## 2. Relative quantitation

During relative quantitation, changes in sample gene expression are measured based on either an external standard or a reference sample, also known as a calibrator (Livak and Schmittgen 2001). When using a calibrator, the results are expressed as a target/reference ratio. There are numerous mathematical models available to calculate the mean normalized gene expression from relative quantitationassays. Depending onthe method employed, these can yield different results and thus discrepant measures of standard error (Liu and Saint 2002; Muller, Janovjak et al. 2002). Table 1 shows a comparison of the different methods, with an explanation of each method to follow (Wong and Medrano 2005).

## 9. Quantitative analyses

For quantitative analysis, the amplification curves are evaluated. The amplification process is monitored either through the fluorescence of dsDNA-specific dyes (like SYBR Green I) or ofsequence-specific probes. Each curve consists of at least three distinct phases: 1) an initial lag phase in which no product accumulation can be measured, 2) an exponential phase, and 3) a plateau phase (Wilhelm and Pingoud 2003). The exponential phase in principle could be extrapolated to the start of the reaction (Cycle 0) to calculate the template copy number, but the error would be too high. The template copy number can be estimated with greater precision from the number of cycles needed for the signal to reach an arbitrary threshold. The threshold must intersect the signal curve in its exponential phase, in which the signal increase correlates with product accumulation. The intersection point is the so-called threshold value ($C_T$) or crossing point ($C_P$). This point may be between two successive cycles (i.e. it may be a fractional number). For exact quantifications, the efficiency of the amplification reaction must be known. It is crucial that the amplification efficiencies of standards and unknowns are identical (Wilhelm and Pingoud 2003). The efficiency can be estimated from the $C_T$ values of samples with known template concentrations ('standards') as described below (Wilhelm and Pingoud 2003).

During the exponential phase, the signal $S$can be described by Equation 1:

$$S = pN_0\varepsilon^c \tag{1}$$

where$p$is a proportionality factor to relate PCR product concentration and signal intensity, $N_0$is the amount of template, $\varepsilon$is the amplification efficiency ($1 \leq \varepsilon \leq 2$; $\varepsilon = 2$ means 100% efficiency) and $c$is the cycle number.

Solving for c results in Equation 2:

$$c = -(\log\varepsilon)^{-1}(\log N_0 + \log p - \log S) \tag{2}$$

With $m = -(\log\varepsilon)^{-1}$ and $b = -(\log\varepsilon)^{-1}(\log p - \log S)$, Equation 2 simplifies to Equation 3:

$$c = m\log N_0 + b \tag{3}$$

This equation describes the linear relationship between the $C_T$values determined and the log of the template concentration ($N_0$). The parameters $m$and $b$can be determined by a regression analysis of the $C_T$ values of the standards. When solved for $N_0$, this equation serves as a calibration curve for the calculation of the unknowns according to Equation 4:

$$N_0 = 10^{(C_T - b)/m} \tag{4}$$

| Methods (Reference) | Amplification Efficiency Correction | Amplification Efficiency Calculation | Amplification Efficiency Assumptions | Automated Excel-Based Program |
|---|---|---|---|---|
| Standard Curve (31) | no | standard curve | no experimental sample variation | no |
| Comparative $C_t$ ($2^{-\Delta\Delta Ct}$) (21) | yes | standard curve | reference = target | no |
| Pfaffl et al. (26) | yes | standard curve | sample = control | REST[a] |
| Q-Gene (23) | yes | standard curve | sample = control | Q-Gene[b] |
| Gentle et al. (7) | yes | raw data | researcher defines log-linear phase | no |
| Liu and Saint (22) | yes | raw data | reference and target genes can have different efficiencies | no |
| DART-PCR (30) | yes | raw data | statistically defined log-linear phase | DART-PCR[c] |

$C_t$, cycle threshold; DART-PCR, data analysis for real-time PCR; REST, relative expression software tool.

[a]www.gene-quantification.info
[b]www.BioTechniques.com
[c]nar.oupjournals.org/cgi/content/full/31/14/e73/DC1

Table 1. Characteristics of Relative Quantitation Methods

| Detection Chemistries | Specificity | Multiplex Capability | Specific Oligonucleotide Required | Allelic Discrimination | Cost |
|---|---|---|---|---|---|
| DNA Binding Dyes | two PCR primers | No | No | No | $ |
| Hybridization Probe Four Oligonucleotide Method | two PCR primers; two specific probes | Yes | Yes | Yes | $$$ |
| Hybridization Probe Three Oligonucleotide Method | two PCR primers; one specific probe | Yes | Yes | Yes | $$$ |
| Hydrolysis Probes | two PCR primers; one specific probe | Yes | Yes | Yes | $$$ |
| Molecular Beacons | two PCR primers; one specific probe | Yes | Yes | Yes | $$$ |
| Scorpions | one PCR primer; one primer/probe | Yes | Yes | Yes | $$$ |
| Sunrise Primers | two PCR primers | Yes | Yes | Yes | $$$ |
| LUX Primers | two PCR primers | Yes | Yes | No | $$ |

$$$, very expensive; $$, moderately expensive; $, inexpensive. LUX, light upon extension.

Table 2. Characteristics of Detection Chemistries

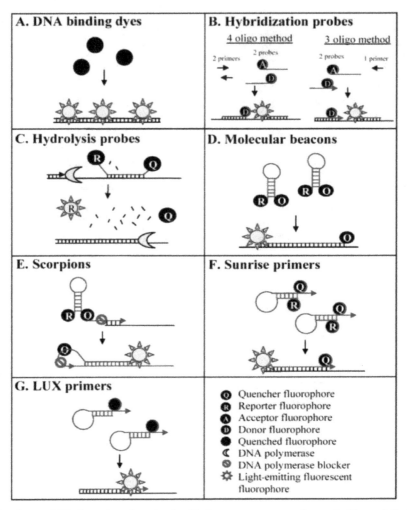

Fig. 1. Real-time PCR detection chemistries. Probe sequences are shown in blue while target DNA sequences are shown in black. Primers are indicated by horizontal arrowheads. Not all unlabeled PCR primers are shown. Oligo, oligonucleotide.

The efficiency can be calculated from the parameter m by using Equation 5:

$$\varepsilon = 10^{-1/m} \qquad (5)$$

By inserting $\varepsilon$ back into Equation 4, one obtains Equation 6:

$$N_0 = \varepsilon^{(b-C_T)} \qquad (6)$$

The maximum value for $\varepsilon$ is 2.0 (i.e. the amount of product is doubled in each cycle). The experimental value for $\varepsilon$ usually varies between 1.5 and 1.9. Lower efficiencies limit the sensitivity of the assay but allow quantifications with higher precisions. Therefore, reactions should be optimized for high efficiency. The effect of the efficiency on the precision, however, is not pronounced.

With more than six orders of magnitude, the dynamic range of this procedure is extraordinarily high (Marcucci, Livak et al. 1998; Verhagen, Willemse et al. 2000;Sails, Fox et al. 2003). The accuracy of this technique is limited by the precision of the determination of the $C_T$ values. The error of the $C_T$ values results from the signal noise and the $C_T$ calculation method. In highly optimized assays, standard errors of less than ± 0.2 cycles can be achieved. By assuming an amplification efficiency of 2 (i.e. 100 %), this implies that the minimum relative error for the quantification is about 10- 20%. The effects of different analysis and calculation methods and the effects of amplification-independent signal trends on the accuracy and precision of quantifications by realtime PCR are described in detail in papers by Lui et al. and Wilhelm et al (Liu and Saint 2002; Wilhelm, Pingoud et al. 2003).

Quantification is relative to the standard used. Only when the absolute concentration of the template molecules in the standard sample is known can the results be absolute. However, in most cases, determination of absolute concentrations is not required. That real-time PCR allows absolute quantification is demonstrated in principle by the reported determination of genome sizes (Wilhelm, Pingoud et al. 2003).

All quantifications by PCR are relative, either to a standard or to a reference gene. Interestingly, Equation 6 nicely illustrates the relative character of the quantifications using a dilution series of a standard; the meaning of the parameter b is the expected $C_T$ value of a sample with 'one' copy (or any other unit as defined by the operator). The difference of this value minus the $C_T$ value determined for the unknown sample ($\Delta C_T = b - C_T$) is a direct measure for the relative difference in template concentrations of the unknown and standard (Wilhelm and Pingoud 2003).

To analyse relative changes in transcript levels, the chosen standard is usually a reference transcript, for example from a housekeeping gene, itself with unknown template concentration. The calculation of $\Delta C_T$ values between reference and sample transcript in a reference and a test sample then provides a simple tool to estimate relative changes. The derivation, assumptions and applications of the so-called $2^{\Delta\Delta CT}$ method are described elsewhere by Livak et al (Livak and Schmittgen 2001). The results of this method are only semiquantitative because the efficiency $\varepsilon$ is assumed to be 2.0 in all experiments and for all templates, which is at best an optimistic estimate. More precise results are obtained with a procedure introduced by Pfaffl et al., which includes a measured value for $\varepsilon$ (Pfaffl 2001; Wilhelm and Pingoud 2003).

In general, care must also be taken for accurate quantifications with external standardization, especially with respect to polymerase inhibitors, which may be present in differentconcentrations in the unknowns and standards. This problem is circumvented by internal standardization. Here, an analytically distinguishable standard template ('competitor') is added to the sample and co-amplified in the same reaction (Gilliland, Perrin et al. 1990; Goerke, Bayer et al. 2001). The direct and simultaneous quantitative analysis of both products in realtime PCR also poses problems. These difficulties are mostly due to the fact that different fluorophors have to be used to distinguish the sequences of competitor and sample. As a result of different FRET and quantum efficiencies, the $C_T$ values obtained for competitor and sample are not directly comparable.

The problem of where to set the threshold makes relativequantifications difficult if not impossible. However, a simple trickcan be used to combine the advantages of both methods: thereaction mixtures are prepared in duplicate (Gibson, Heid et al. 1996). To one of thesemixtures, the probe specific for the competitor sequence isadded, whereas the probe specific for the sample sequence isadded to the other mixture. This process is carried out for a seriesof reactions with different amounts of competitor added. Withthis procedure, two calibration lines are obtained and theintersection of the two lines is the equivalence point (Wilhelm and Pingoud 2003).

## 10. Melting curve analyses

Melting curves represent the temperature dependence of the fluorescence. They are recorded subsequent to the amplification of the target sequence by PCR. The detection can be performed either with dsDNA-specific dyes like SYBR Green I or with sequence-specific probes such as the molecular beacons and the hybridisation probes (scorpion and sunrise primers cannot be used for melting curve analysis because they are integrated into the PCR products; TaqMan probes cannot be used for melting curve analyses either, since their signal generation depends on the hydrolysis of the probe). Melting curves of sequence-specific probes are used for genotyping, resolving single base mismatches between target sequence and probe (Lay and Wittwer 1997; Whitcombe, Brownie et al. 1998), whereas SYBR Green I is used most frequently for product characterization (Ririe, Rasmussen et al. 1997). It has been reported that melting curves measured with SYBR Green I can also be utilized for genotyping of insertion/deletion polymorphisms and of single nucleotide polymorphisms (SNPs) (Akey, Sosnoski et al. 2001; Lin, Tseng et al. 2001).

In melting curves, the signal decreases gradually as a result of a temperature-dependent quench and more abruptly at a certain temperature because of the melting of the products

(dsDNA or ssDNA/probe hybrid). The melting temperature ($T_m$) of a product is defined as the temperature at which the steepest decrease of signal occurs. This can be identified conveniently as the peak value(s) (global or local maxima) in the negative derivative of the melting curve. Additionally, the area under the curve (AUC) of the peaks is proportional to the amount of product. Therefore, melting curve analysis may be used for quantifications with internal standardization when the $T_m$ values of sample and competitor products are significantly different (Al-Robaiy, Rupf et al. 2001). However, well-performed normalization is required to reduce the systematic error due to the temperature dependent quench. This quench also limits the sensitivity of melting curve analyses. At present, there is only one software package available that can remove the quench effects from the data (Wilhelm, Pingoud et al. 2003).

With SYBR Green I, the amplification of the correct target sequence can be confirmed. In most cases, nonspecific products have different lengths and therefore deviating melting temperatures (Ririe, Rasmussen et al. 1997).Hybridisation probes, molecular beacons and TaqMan probes are used for mutation detection (Lay and Wittwer 1997; Bernard, Ajioka et al. 1998; Bernard and Wittwer 2000), genotyping (Whitcombe, Brownie et al. 1998; Ulvik and Ueland 2001; Grant, Steinlicht et al. 2002; Randen, Sørensen et al. 2003) and SNP screening (Sasvari-Szekely, Gerstner et al. 2000; Mhlanga and Malmberg 2001).

## 11. Applications

Real-time PCR is used for absolute and relative quantifications of DNA and RNA template molecules and for genotyping in a variety of applications (Wilhelm and Pingoud 2003).

Quantitative real-time PCR is used to determine viral loads (Mackay, Arden et al. 2002),gene expression (Bustin 2000; Goerke, Bayer et al. 2001), titers of germs and contaminations (infood, blood, other body fluids and tissues) (Locatelli, Urso et al. 2000; Hernandez, Rio et al. 2001; Norton 2002), allele imbalances (Ruiz-Ponte, Loidi et al. 2000) and the degrees of amplification and deletion ofgenes (Chiang, Wei et al. 1999; Nigro, Takahashi et al. 2001).

Real-time PCR is also becoming increasingly important in thediagnosis of tumors, such as for the detection and monitoringof minimal residual diseases (Marcucci, Livak et al. 1998; Elmaagacli, Beelen et al. 2000; Amabile, Giannini et al. 2001; Krauter, Heil et al. 2001; Krauter, Hoellge et al. 2001), the identification of micrometastases in colorectal cancer (Bustin, Gyselman et al. 1999), neuroblastoma (Cheung and Cheung 2001) and prostate cancer (Gelmini, Tricarico et al. 2001). It has been used to quantify amplifications of oncogenes (Bieche, Laurendeau et al. 1999; Lehmann, Glöckner et al. 2000; Lyon, Millson et al. 2001; Konigshoff, Wilhelm et al. 2003) as well as deletions of tumor suppressor genes in tumor samples (Wilhelm and Pingoud 2003). Also, the response of human cancer to drugs has been studied (Au, Chim et al. 2002; Miyoshi, Ando et al. 2002;Reimer, Koczan et al. 2002). Other clinically relevant applications are cytokine mRNA profiling in immune response (Hempel, Smith et al. 2002; Stordeur, Poulin et al. 2002) and tissue-specific gene expression analysis (Bustin 2002; Poola 2003; Prieto-Alamo, Cabrera-Luque et al. 2003).

Also, the results of DNA chip experiments are validated by real-time PCR quantifications (Miyazato, Ueno et al. 2001; Rickman, Bobek et al. 2001;Crnogorac-Jurcevic, Efthimiou et al. 2002).

Chimerism analysis is possible when sequence-specific probes are utilized to differentiate and quantify alleles. High dynamic ranges can be achieved with allele-specific real-time PCR (Shively, Chang et al. 2003). Robust chimerism analyses with extremely large dynamic ranges based on insertion/deletion polymorphisms and on SNPs are also possible (Wilhelm, Reuter et al. 2002; Maas, Schaap et al. 2003). Genetic chimerisms have been monitored by Y-chromosome-specific real-time PCR for sex-mismatched transplantations (Fehse, Chukhlovin et al. 2001; Byrne, Huang et al. 2002; Elmaagacli 2002) and by allele-specific real-time PCR (Maas, Schaap et al. 2003; Shively, Chang et al. 2003). This combination of allele-specific amplification with real-time PCR has been shown to reveal detection limits of down to 0.01% for SNPs (Maas, Schaap et al. 2003). Real-time PCR is increasingly used in forensic analyses (Andreasson, Gyllensten et al. 2002; von Wurmb-Schwark, Higuchi et al. 2002; Ye, Parra et al. 2002), but also to monitor disease- or age-related accumulation of deletions in the mitochondrial genome (Mehmet, Ahmed et al. 2001; He, Chinnery et al. 2002).

Melting curve analyses are used for real-time competitive PCR (Al-Robaiy, Rupf et al. 2001; Lyon, Millson et al. 2001), gene dosage tests (Ruiz-Ponte, Loidi et al. 2000) and genotyping and SNP detection (Bullock, Bruns et al. 2002; Burian, Grosch et al. 2002; Randen, Sørensen et al. 2003). These applications will have a particularly strong impact on pharmacogenetics (Palladino, Kay et al. 2003). Profiling of DNA methylation is also possible by melting curve analysis (Worm, Aggerholm et al. 2001; Akey, Akey et al. 2002), which simplifies the analysis of epigenetic variations of the genome and developmental processes.

In brief, the advantages of real-time PCR are exploited in clinical diagnosis and the monitoring of infectious diseases and tumors. The technique is applied for the analysis of age dependent diseases, cytokine and tissue-specific expression, forensic samples, epigenetic factors like DNA methylation and for food monitoring. The field of applications is still growing rapidly, which suggests that real-time PCR will become one of the most important techniques in molecular life sciences and medicine (Wilhelm and Pingoud 2003).

## 12. Normalization

Gene expression analysis at the messenger RNA (mRNA) level has become increasingly important in biological research. Generally we detect RNAs to determine if differences protein expression could be explained at the transcriptional level. In particular, measurement of mRNA is needed in situations where quantification of the protein is difficult or cumbersome. Most recently, mRNA expression analysis is being used to provide insight into complex regulatory networks and to identify genes relevant to new biological processes or implicated in diseases (Hendriks-Balk, Michel et al. 2007).Common methods for RNA detection include: Northern blotting, in situ hybridization, qualitative RTPCR, RNase protection assay, competitive RT-PCR, microarray analysis, and quantitative real-time PCR. The specificity, wide dynamic range , ease-of-use , requiring a minimal amount of RNA, no post-PCR handling and avoiding the use of radioactivity, has made the real-time quantitative reverse transcription polymerase chain reaction (qRT-PCR) the method of choice for quantitating RNA levels (Radonic, Thulke et al. 2004). The technique has two main steps: CDNA synthesis by reverse transcription of mRNA and subsequent quantification of specific CDNAs by real-time PCR. It is in many cases the only method for measuring mRNA levels of vivo low copy number targets of interest for which alternative assays either do not exist or lack the required sensitivity so these specification has led to made it the "gold standard" for mRNA quantification (Huggett, Dheda et al. 2005). Most gene expression assays are based on the comparison of two or more samples and require uniform sampling conditions for this comparison to be valid. Unfortunately, many factors can contribute to variability in the analysis of samples, making the results difficult to reproduce between experiments. During the preparation of CDNA for real-time PCR analysis there is significant potential for small errors to accumulate. For example, differences in sample size, RNA extraction efficiency, pippetting accuracy and reverse transcription efficiency will all add variability to your samples (Huggett, Dheda et al. 2005). Not only can the quantity and quality of RNA extracted from multiple samples vary, but even replicates can vary dramatically due to factors such as sample degradation, extraction efficiency, and contamination. On the other hand, since many biological samples contain inhibitors of the RT and/or the PCR step, it is crucial to assess the presence of any inhibitors of polymerase activity in RT and PCR. so it is clear that we need to incorporate some normalization method to control for errors. The identification of a valid reference for data normalisation remains the most stubborn of problems and none of the solutions proposed are ideal. Normalization methods range from ensuring that a similar sample size is chosen to using an internal housekeeping or reference gene (Table 3) (Huggett, Dheda et al. 2005).

| Normalisation strategy | Pros | Cons | Note |
|---|---|---|---|
| Similar sample size/tissue volume | Relatively easy | Sample size/tissue volume may be difficult to estimate and/or may not be biological representative | Simple first step to reduce experimental error |
| Total RNA | Ensures similar reverse transcriptase input. May provide information on the integrity (depending on technique used) | Does not control for error introduced at the reverse transcription or PCR stages. Assumes no variation in rRNA/mRNA ratio | Requires a good method of assessing quality and quantity |
| Genomic DNA | Give an idea of the cellular sample size. | May vary in copy number per cell. Difficult to extract with RNA | Rarely used. Can be measured optically or by real time PCR |
| Reference genes ribosomal RNAs (rRNA) | Internal control that is subject to the same conditions as the RNA of interest. Also measured by real time RT-PCR | Must be validated using the same experimental samples. Resolution of assay is defined by the error of the reference gene | Oligo dt priming of RNA for reverse transcription will not work well with rRNA as no polyA tail is present. Usually in high abundance |
| Reference genes messenger RNAs (mRNA) | Internal control that is subject to the same conditions as the mRNA of interest. Also measured by real time RT-PCR | Must be validated using the same experimental samples. Resolution of assay is defined by the error of the reference gene | Most, but not all, of mRNAs contain polyA tails and can be primed with oligo dt for reverse transcription |
| Alien molecules | Internal control that is subject to most of the conditions as the mRNA of interest. Is without the biological variability of a reference gene | Must be identified and cloned or synthesised. Unlike the RNA of interest, is not extracted from the within the cells | Requires more characterisation and to be made available commercially |

There is good correlation between the RNA concentration used and the real time PCR estimation of the different amounts of HuPO cDNA (using omniscript reverse transcriptase).

(Huggett, Dheda et al. 2005)

Table 3. Comparison of the actual amount of RNA used in different reverse transcription reactions with the respective amount of HuPO
Comparison of the actual amount of RNA used in different reverse transcription reactions with the respective amount of HuPO
cDNA measured by real-time RT-PCR

## 12.1 Methods of normalization

### 1. Standardizing Sample size

The most basic method of normalization ensures that an experiment compares similar sample sizes and this is achieved by measuring tissue weight, volume or cell number. This method can reduce the experimental error of first stage of qRT-PCR. It seems to be straightforward, but we can't ensure that equal volume of different samples contain the same cellular material. Real-time RT-PCR experiments that rely on the extraction of RNA from complex tissue samples are averaging the data from numerous, variable subpopulations of cells of different lineage at different stages of differentiation (Bustin, Benes et al. 2005). This can be misleading, as is illustrated when sampling a similar volume of blood from HIV +ve patients. Patients with HIV that have less advanced immunosupression (CD4 counts X200 cells/ml) will yield a higher amount RNA than patients with CD4 counts p200 cells/ml. This is simply because there are fewer cells per milliliter of blood in the latter group. Even cellular subpopulations of the same pathological origin can be highly heterogeneous. Tumor biopsies, in particular, are made up not just of normal and cancer epithelial cells, but there may be several subclones of epithelial cancer cells together with stromal, immune and vascular components (Vandesompele, De Preter et al. 2002; Bustin, Benes et al. 2005). This variability can give us misleading or meaningless result to solve this we can use laser capture microdissection to normalize against the dissected area which can report the target mRNA levels conveniently as copies per area or cell dissected. In in vitro cell culture, due to different morphologies or clumping up of cells, it's hard to determine sample size (cell number) .we can treat them with buffers and/or enzymes till they could be counted, however these treatments surely could affect gene expression. This approach could not be applied for solid tumors for which the amount of cells cannot be determined accurately. To work around this problem, it was suggested to standardize the RT-qPCR data between samples using the amount of genomic DNA as an indicator reflecting the number of cells in each sample. However, these approaches do not account for the degradation of RNA or the efficiency of RT and PCR. So, while ensuring a similar sample size is important it clearly is not sufficient on its own (Huggett, Dheda et al. 2005).

## 2. Normalization with genomic DNA

Another method for normalization is measuring the amount of genomic DNA (gDNA). This appears to be an ideal method as it does not require reverse transcription for detection by real-time PCR (Bustin 2002). However, this approach do not account for the degradation of RNA or the efficiency of RT and PCR. .Moreover, in the case of normalization with genomic DNA, the fact of working with tumor cells can present additional problems because they tend to have abnormal karyotype. Therefore, the ratio between the amount of DNA and cell number is variable (Huggett, Dheda et al. 2005). Another major problem with using this strategy is that RNA extraction procedures are usually not designed to purify DNA, so the extraction rate may vary between different samples, with DNA yields often being low. In conclude Normalization against genomic DNA is rarely used since it is difficult to coextract with RNA and it may vary in copy number per cell (Huggett, Dheda et al. 2005).

## 3. Normalization with total RNA

The normalization of RT-qPCR results can be compared to the amount of total RNA used in the reverse transcription step. Not only does this facilitates normalization but circumvents problems associated with the linearity of the reverse transcriptase step. There are several methods for quantifying RNA; the most common is to measure the absorbance at 260 nm ($A_{260}$) with a UV spectrophotometer. The major advantage of this spectrometer, whose sensitivity is estimated at 5 ng/uL, is that it requires only 1 microL of sample, placed in direct contact with the optical system (Huggett, Dheda et al. 2005; Hendriks-Balk, Michel et al. 2007). However, contaminants absorbing at 260 nm, such as proteins, phenol or genomic DNA, can lead to overestimated results. Another optical system is flourimetry in which intercalating fluorescent nucleic acid is used, the kit RiboGreen ® Molecular Probes based on this principle. This is a more sensitive technique but does not discriminate RNA from DNA, and contaminants such as phenol can produce variable results (Huggett, Dheda et al. 2005). Since it is generally assumed that OD260 analysis is less accurate than the RiboGreen assay, we have compared RNA quantification data obtained using the RiboGreen assay with

OD260 analysis using a Genequant II (Pharmacia). The results (Fig. 2) suggest that both methods generate comparable results when the RNA concentration is not less than 100 ng/μl, with RiboGreen measurements lower than those obtained using the spectrophotometer. OD260 analysis becomes less reliable at lower RNA concentrations (Bustin 2002).

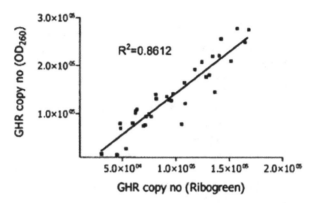

Fig. 2. Comparison of RNA quantification using the Genequant II and the RiboGreen fluorescent assay. RNA from 34 normal colon biopsies was quantitated using a standard Genequant II protocol which measures the absorbance at 260 nm. The same samples were then quantitated using a standard RiboGreen fluorescent assay. RT-PCR assays targeting the GHR were carried out and Ct normalised against the respective concentrations determined by the two methods. The scatterplot shows a good correlation between the two methods ($r2=0.8612$) (Bustin 2002).

What is also important, but often overlooked, is the need to assess the quality of RNA because degraded RNAs may adversely affect results. The opportune development of Agilent's 2100 Bioanalyser and LabChip technology has provided a new standard of RNA quality control as well as permitting concomitant quantification of RNA. The analysis is not influenced by contamination of phenol or proteins, against the presence of genomic DNA requires a correction of the measurement of the concentration of RNA (Bustin, Benes et al. 2005). This is particularly important when the RNA has been extracted from 'dirty' tissue such as the colon (Vandesompele, De Preter et al. 2002). This technique allows characterizing the RNA in a concentration between 5 and 500 ng/uL. For each sample, the software determines the ratio 28S/18S and assigns a RIN (RNA Integrity Number) which takes into account the entire electropherogram. The value of RIN ranges from 1 to 10, with 1 being totally degraded RNA, and 10 to a high-quality RNA. In addition to being fast and allow high throughput, this technique requires only 1 microL of sample. It is the simplest method and objective qualitative analysis of RNA, its use is recommended (Schmittgen and Zakrajsek 2000; Bustin 2002). Similar in concept, but requiring an additional RT-PCR assay, is normalization against one of the rRNAs. rRNA levels may vary less under conditions that affect the expression of mRNAs and the use of rRNA has been claimed to be more reliable than that of several reference genes in rat livers and human skin fibroblasts (Bustin 2002; Huggett, Dheda et al. 2005). But, a drawback is that it primarily measures ribosomal RNA (rRNA) whereas real-time PCR aims to determine mRNA expression and normalization for total RNA assumes that the rRNA:mRNA ratio is the same in all groups, which might not always be the case. Moreover, rRNA is not present in purified mRNA and the high abundance of rRNA compared to mRNA makes it difficult to subtract the baseline value in realtime PCR analysis. Thus, markers of rRNA such as 18S or 28S rRNA might also be suboptimal as normalization factors in many settings (Hendriks-Balk, Michel et al. 2007). Also, it has been reported that rRNA transcription is affected by biological factors and drugs. An important parameter to consider when normalized relative to the RNA is the quality of it. Differences in the quality of the samples strongly depend on the extraction step and are the source of the most common variations in RT-qPCR. It is therefore important to use the same method of extraction for all samples analyzed (Vandesompele, De Preter et al. 2002). The quality of RNA is defined by both its purity (no contamination) and integrity (non-degraded RNA). Its purity was determined by measuring the absorbance at 230 (organic contaminants) and 280 nm (specific proteins), RNA is considered pure ratios $A_{260}/A_{230}$ and $A_{260}/A_{280} > 1.8$. With regard to the integrity, the traditional method was to visualize the bands of 28S and 18S ribosomal RNA on a gel electrophoresis. Indeed, it is difficult to analyze directly the mRNA; they represent only 1% to 3% of total RNA. We must therefore consider that the degradation of ribosomal RNA, the majority, reflecting the degradation of mRNA. Thus, the 18S/28S ratio assesses the integrity of RNA; a ratio close to 2 is considered an indicator of RNA with little or no gradient. However, this method requires a large amount of RNA (0.5-2 mg), and is not sensitive enough to detect slight damage. Normalizing a sample against total RNA has the drawback of not controlling for variation inherent in the reverse transcription or PCR reactions and it ignores the efficiency of converting RNA into CDNA. Also rRNA cannot be used for normalization when quantifying targets from polyA-enriched samples (Huggett, Dheda et al. 2005). A final drawback when using total RNA for normalization is the lack of internal control for RT or PCR inhibitors. All quantitative methods assume that the RNA targets are reverse transcribed and subsequently amplified with similar efficiency but the reaction is extremely sensitive to the presence of inhibitors, which can be reagents used in the extraction step (salts, alcohols, phenols), or components copurification organic (urea, heme, heparin, immunoglobulin G). These compounds can also inhibit the PCR reaction. Thus, two reactions with an equal amount of RNA, but the efficiencies of RT and / or PCR are different, will yield results that cannot be compared. Different methods exist to assess the presence of inhibitors in biological samples. First, it is possible to compare the efficacy of PCR for different dilutions (1/20 and 1/80 for example) of a sample. An alternative is to add a defined amount of a synthetic single-stranded amplicon CDNA samples, and comparing its amplification compared to a control without CDNA. However, these methods are limited to verify the absence of PCR inhibitors, and do not evaluate the effectiveness of the reverse transcription step (Bustin 2002).

### 4. Normalization with an artificial molecule (spike)

An interesting solution to control the two enzymatic reactions (RT and PCR) is added to RNA extracted an exogenous RNA, which will compare the amplification between the different samples. This sequence control should show no similarity to the target RNA, we will use such a specific mRNA from a plant when studying gene expression in humans. The main criticism of using spikes is that, while they can be introduced prior to extraction, unlike the cellular RNAs they are not extracted from within the tissue. Consequently, there may be situations (e.g. if the samples differ histologically) when the spike may not be a good control for the extraction procedure. The stages required to generate the alien molecule may also not be feasible for small laboratories wanting to perform limited amounts of real-time RTPCR (Schmittgen and Zakrajsek 2000; Argyropoulos, Psallida et al. 2006).

## 5. Normalization with reference genes

Reference genes represent the by far most common method for normalizing qRT-PCR data. Reference genes are often referred to as housekeeping genes assuming that those genes are expressed at a constant level in various tissues at all stages of development and are unaffected by the experimental treatment (Hendriks-Balk, Michel et al. 2007; Balogh, Paragh et al. 2008). Use of this endogenous control theory allows controlling all stages of the experimental protocol; its expression reflects not only the quantity and quality of RNA used, but the efficiencies of the RT and PCR. An advantage of reference genes as compared to total or rRNA is that the reference gene is subject to the same conditions as the mRNA of interest (Hendriks-Balk, Michel et al. 2007). The most commonly used reference genes include β-actin (ACTB), (GAPDH), (HPRT) and 18S rRNA. The other commonly used reference genes would be PGK1, B2M, GAPD, HMBS, HPRT1, RPL13A, SDHA, TBP, UBC and YWHAZ (Vandesompele, De Preter et al. 2002). The initial concentration of a target is usually derived from the $C_T$ (cycle threshold), which is the number of amplification cycles where the amplification curve crosses the threshold line. This line is placed at the exponential phase, so as to be clearly distinguishable from background noise. For each sample, the $C_T$ obtained for the genes of interest and reference must be converted to normalized expression ratio. For this, various options are available, they are integrated in the software provided with the various qPCR instruments or described in the literature. The relative standard curve method requires the construction, for the target gene and reference gene, a range made from a series of dilutions of a reference sample. These ranges to obtain standard curves, obtained by expressing the $C_T$ as a function of log of the initial concentration of cDNA. Concentration values for each point of the range can be set arbitrarily in accordance with the dilution factors. Therefore, the relative amount of a target is determined by the $C_T$ by interpolation with the standard curve. The standard expression of a gene of interest is determined by the following formula:

$$R = \frac{Relative\ amount\ of\ the\ gene\ of\ intrest}{Relative\ amount\ of\ the\ refrence\ gene}$$

In addition, a calibrator is typically used. This is a sample used as a basis for comparing results. The normalized ratio of each sample is divided by the normalized ratio of the calibrator. Thus, the calibrator becomes the reference 1x, and all other samples are expressed as a ratio relative to the calibrator. The method of $\Delta\Delta C_T$ uses a mathematical formule to calculate the ratio of expression of a target gene between two samples, normalized with reference gene. First, the differences $\Delta C_T$ between the values of $C_T$ target gene and reference gene were determined for the test sample and control.

$$\Delta C_{T\ (sample)} = C_T\ (target\ sample) - C_T\ (reference\ sample)$$

$$\Delta C_{t\ (control)} = C_T\ (target\ control) - C_T\ (reference\ control)$$

Next, the $\Delta\Delta C_T$ between control and the sample is calculated:

$$\Delta\Delta C_T \, \Delta C_T \,_{(control)} - \Delta C_T \,_{(sample)}$$

Finally, the normalized ratio of expression of a target gene is determined by the formula: $2^{-\Delta\Delta CT}$.

Unlike the relative standard curve method, where the amplification efficiency (E) target genes and reference is directly taken into account when building ranges, the method of $\Delta\Delta C_T$ is assumed that the efficiencies of the two genes are equal to 100% (E = 2, with each cycle of the exponential phase, the concentration of PCR products is doubled). However, a difference in PCR efficiency of 3% ($\Delta E = 0.03$) between the two genes results in an error of 47% for the ratio of expression if $E_{target} < E_{ref}$ and 209% if $E_{target} > E_{ref}$ after 25 cycles. In addition, the error increases exponentially with larger variations of efficiency and a greater number of cycles. New models have been developed taking into account the efficiency of PCR target gene and reference gene. The most common is the model of Pfaffl, where the relative expression ratio (R) of a target gene between a sample and control is determined by the following formula:

$$R = \frac{(E_{target})^{\Delta C \, Ttarget(control \, -sample)}}{(E_{reference})^{\Delta C \, Treference(control \, -sample)}}$$

In this model of Pfaffl, the efficiency of PCR for a given gene is calculated from the construction of a calibration curve using the following formula: $E = 10^{[-1/ \, gradient]}$. This method gives a good estimate of effectiveness, although it is possible that it is overestimated. However, this approach assumes that the amplification efficiencies between the diluted samples are identical, creating a linear relationship between $C_T$ and amount of CDNA in the beginning. Therefore, some authors such as Liu and Saint have developed models that take into account standards of efficiency for each sample, the latter being determined by the kinetics of the amplification curve. However, with this kind of approach, the slightest error in the measurement of effectiveness is amplified and passed exponentially on the expression ratio calculated. The different models of normalization with reference genes therefore have all the advantages and disadvantages. At present, there is no time-honored method for the treatment of the results of RT-qPCR. Normalization to a reference gene is a simple method and frequently used because it can control many variables. An advantage of reference genes as compared to total or rRNA is that the reference gene is subject to the same conditions as the mRNA of interest (Bustin, Benes et al. 2005; Hendriks-Balk, Michel et al. 2007). What has become apparent over recent years is that there is no single reference gene for all experimental systems. Quantified errors related to the use of a single reference gene as more than three-fold in 25% and more than six-fold in 10% of samples. Today it is clear that reference genes must be carefully validated for each experimental situation and those new experimental conditions or different tissue samples require re-validation of the chosen reference genes (Balogh, Paragh et al. 2008). If inappropriate reference genes are used for normalization, the experimental results obtained can differ greatly from those using a validated reference gene. Validation of a reference gene requires removal of any non-specific variation in expression. This can be done using a recently introduced program called geNorm (freely available at http://medgen.ugent.be/~jvdesomp/genorm/) that mathematically identifies the most suitable reference gene for a given experimental condition. Because of the inherent variation in the expression of reference genes the use of multiple reference genes rather than one reference gene is recommended to ensure reliable normalization of real-time PCR. Several statistical programs can help to determine the most appropriate reference gene or set of genes (Hendriks-Balk, Michel et al. 2007; Borges, Ferreira et al. 2010).

## 13. Abbreviation

ACTB: Beta actin
B2M: Beta-2-microglobulin
$C_T$: Threshold Cycle
FRET: Fluorescence Resonance Energy Transfer
GAPD: Glyceraldehyde-3- phosphate dehydrogenase
HMBS: Hydroxymethyl-bilane synthase
HPRT: hypoxanthine ribosyltransferase
HPRT1: Hypoxanthine phosphoribosyl-transferase 1
HRM: High Resolution Melting
KPCR: Kinetic Polymerase Chain Reaction
LNA: Locked Nucleic Acid
LUX: Light Upon Extension
PCR: Polymerase Chain Reaction
PGK1: phosphoglycerokinase 1
PNA: Peptide Nucleic Acid
Q-PCR (QRT-PCR): Quantitative Real-Time Polymerase Chain Reaction
RPL13A: Ribosomal protein L13a
RT PCR: Reverse transcription PCR
SDHA: Succinate dehydrogenase complex, subunit A
TBP: TATA box binding protein
$T_m$: Melting Point
UBC: Ubiquitin C
YWHAZ: Tyrosine 3-monooxygenase/ tryptophan 5-monooxygenase activation protein, zeta polypeptide

## 14. References

Ahmad, A. I. (2007). "BOXTO as a real-time thermal cycling reporter dye." *Journal of Biosciences*32(2): 229-239.

Akey, D. T., J. M. Akey, et al. (2002). "Assaying DNA methylation based on high-throughput melting curve approaches." *Genomics*80(4): 376-384.

Akey, J., D. Sosnoski, et al. (2001). "Research Report Melting Curve Analysis of SNPs (McSNP®): A Gel-Free and Inexpensive Approach for SNP Genotyping." *Biotechniques*30(2): 358-367.

Al-Robaiy, S., S. Rupf, et al. (2001). "Research Report Rapid Competitive PCR Using Melting Curve Analysis for DNA Quantification." *Biotechniques*31(6): 1382-1388.

Amabile, M., B. Giannini, et al. (2001). "Real-time quantification of different types of bcr-abl transcript in chronic myeloid leukemia." *Haematologica*86(3): 252-259.

Andreasson, H., U. Gyllensten, et al. (2002). "Real-time DNA quantification of nuclear and mitochondrial DNA in forensic analysis." *Biotechniques*33(2): 402-411.

Au, W. Y., C. S. Chim, et al. (2002). "Real-time quantification of multidrug resistance-1 gene expression in relapsed acute promyelocytic leukemia treated with arsenic trioxide." *Haematologica*87(10): 1109-1111.

Ausubel, F. M., R. Brent, et al. (2005). "Current Protocols in Molecular Biology." *Hoboken, NJ: Wiley.*

Bengtsson, M., H. J. Karlsson, et al. (2003). "A new minor groove binding asymmetric cyanine reporter dye for real-time PCR." *Nucleic Acids Res*31(8): e45.

Bernard, P. S., R. S. Ajioka, et al. (1998). "Homogeneous multiplex genotyping of hemochromatosis mutations with fluorescent hybridization probes." *The American journal of pathology*153(4): 1055-1061.

Bernard, P. S. and C. T. Wittwer (2000). "Homogeneous amplification and variant detection by fluorescent hybridization probes." *Clinical chemistry*46(2): 147-148.

Bieche, I., I. Laurendeau, et al. (1999). "Quantitation of MYC gene expression in sporadic breast tumors with a real-time reverse transcription-PCR assay." *Cancer Res*59(12): 2759-2765.

Bio-Rad Laboratories, I. n. c. (2006). "Real-Time PCR Applications Guide."

BioInformatics (2003). "The Market for Real-Time PCR Reagents & Instrumentation." *Arlington VA: BioInformatics.*

Bonnet, G., S. Tyagi, et al. (1999). "Thermodynamic basis of the enhanced specificity of structured DNA probes." *Proc Natl Acad Sci*USA 96(5): 6171-6176.

Braasch, D. A. and D. R. Corey (2001). "Locked nucleic acid (LNA): fine-tuning the recognition of DNA and RNA." *Chem Biol*8(1): 1-7.

Brown, P. O. and D. Botstein (1999). "Exploring the new world of the genome with DNA microarrays." *Nat Genet*21: 33-37.

Bullock, G. C., D. E. Bruns, et al. (2002). "Hepatitis C genotype determination by melting curve analysis with a single set of fluorescence resonance energy transfer probes." *Clinical chemistry*48(12): 2147-2154.

Burian, M., S. Grosch, et al. (2002). "Validation of a new fluorogenic real-time PCR assay for detection of CYP2C9 allelic variants and CYP2C9 allelic distribution in a German population." *Br J Clin Pharmacol*54(5): 518-521.

Bustin, S., V. Gyselman, et al. (1999). "Detection of cytokeratins 19/20 and guanylyl cyclase C in peripheral blood of colorectal cancer patients." *British Journal of Cancer*79(11/12): 1813-1820.

Bustin, S. A. (2000). "Absolute quantification of mRNA using real-time reverse transcription polymerase chain reaction assays." *J Mol Endocrinol* 25: 169-193.

Bustin, S. A. (2002). "Quantification of mRNA using real-time reverse transcription PCR (RT-PCR): trends and problems." *J Mol Endocrinol*29(1): 23-39.

Bustin, S. A. and T. Nolan (2004). "Pitfalls of quantitative real-time reverse-trasncription polymerase chain reaction." *J Biomol Tech*15: 155-166.

Byrne, P., W. Huang, et al. (2002). "Chimerism analysis in sex-mismatched murine transplantation using quantitative real-time PCR." *Biotechniques*32(2): 279-286.

Caplin, B. E., R. P. Rasmussen, et al. (1999). "LightCyclerTM hybridization probes- the most direct way to monitor PCR amplification and mutation detection." *Biochemica*1: 5-8.

Cheung, I. Y. and N. K. Cheung (2001). "Quantitation of marrow disease in neuroblastoma by real-time reverse transcription-PCR." *Clin Cancer Res*7(6): 1698-1705.

Chiang, P. W., W. L. Wei, et al. (1999). "A fluorescent quantitative PCR approach to map gene deletions in the Drosophila genome." *Genetics*153(3): 1313-1316.

Costa, J. M., P. Ernault, et al. (2004). "Chimeric LNA/DNA probes as a detection system for real-time PCR." *Clin Biochem*37(10): 930-932.

Crnogorac-Jurcevic, T., E. Efthimiou, et al. (2002). "Expression profiling of microdissected pancreatic adenocarcinomas." *Oncogene*21(29): 4587-4594.

Crockett, A. O. and C. T. Wittwer (2001). "Fluorescein-labeled oligonucleotides for real-time pcr: using the inherent quenching of deoxyguanosine nucleotides." *Anal Biochem*290(1): 89-97.

Egholm, M., O. Buchardt, et al. (1992). "Peptide nucleic acids (PNA). Oligonucleotide analogs with an achiral peptide backbone." *J Am Chem Soc*114: 1895-1897.

Elmaagacli, A., D. Beelen, et al. (2000). "The amount of BCR-ABL fusion transcripts detected by the real-time quantitative polymerase chain reaction method in patients with Philadelphia chromosome positive chronic myeloid leukemia correlates with the disease stage." *Annals of hematology*79(8): 424-431.

Elmaagacli, A. H. (2002). "Real-time PCR for monitoring minimal residual disease and chimerism in patients after allogeneic transplantation." *Int J Hematol*76 Suppl 2: 204-205.

Espy, M. J., J. R. Uhl, et al. (2006). "Real-time PCR in clinical microbiology: applications for routine laboratory testing." *Clin Microbiol Rev*19(1): 165-256.

Fehse, B., A. Chukhlovin, et al. (2001). "Real-time quantitative Y chromosome-specific PCR (QYCS-PCR) for monitoring hematopoietic chimerism after sex-mismatched allogeneic stem cell transplantation." *J Hematother Stem Cell Res*10(3): 419-425.

Fronhoffs, S., G. Totzke, et al. (2002). "A method for the rapid construction of cRNA standard curves in quantitative real-time reverse transcription polymerase chain reaction." *Molecular and cellular probes*16(2): 99-110.

Gelmini, S., C. Tricarico, et al. (2001). "Real-Time quantitative reverse transcriptase-polymerase chain reaction (RT-PCR) for the measurement of prostate-specific antigen mRNA in the peripheral blood of patients with prostate carcinoma using the taqman detection system." *Clin Chem Lab Med*39(5): 385-391.

Gerard, C. J., K. Olsson, et al. (1998). "Improved quantitation of minimal residual disease in multiple myeloma using real-time polymerase chain reaction and plasmid-DNA complementarity determining region III standards." *Cancer research*58(17): 3957-3964.

Gibson, U. E., C. A. Heid, et al. (1996). "A novel method for real time quantitative RT-PCR." *Genome Res*6(10): 995-1001.

Gilliland, G., S. Perrin, et al. (1990). "Analysis of cytokine mRNA and DNA: detection and quantitation by competitive polymerase chain reaction." *Proceedings of the National Academy of Sciences*87(7): 2725-2729.

Giulietti, A., L. Overbergh, et al. (2001). "An overview of real-time quantitative PCR: applications to quantify cytokine gene expression." *Methods*25(4): 386-401.

Goerke, C., M. G. Bayer, et al. (2001). "Quantification of bacterial transcripts during infection using competitive reverse transcription-PCR (RT-PCR) and LightCycler RT-PCR." *Clinical and Vaccine Immunology*8(2): 279-282.

Grant, S. F. A., S. Steinlicht, et al. (2002). "SNP genotyping on a genome-wide amplified DOP-PCR template." *Nucleic Acids Research*30(22): e125-e125.

He, L., P. F. Chinnery, et al. (2002). "Detection and quantification of mitochondrial DNA deletions in individual cells by real-time PCR." *Nucleic Acids Res*30(14): e68.

Heid, C. A., J. Stevens, et al. (1996). "Real time quantitative PCR." *Genome Res*6(10): 986-994.

Hempel, D. M., K. A. Smith, et al. (2002). "Analysis of cellular immune responses in the peripheral blood of mice using real-time RT-PCR." *J Immunol Methods*259(1-2): 129-138.

Hernandez, M., A. Rio, et al. (2001). "A rapeseed-specific gene, acetyl-CoA carboxylase, can be used as a reference for qualitative and real-time quantitative PCR detection of transgenes from mixed food samples." *J Agric Food Chem*49(8): 3622-3627.

Higuchi, R., G. Dollinger, et al. (1992). "Simultaneous amplification and detection of specific DNA sequences." *Biotechnology*10: 413-417.

Higuchi, R., C. Fockler, et al. (1993). "Kinetic PCR analysis: real-time monitoring of DNA amplification reactions." *Biotechnology (N Y)*11(9): 1026-1030.

Holland, P. M., R. D. Abramson, et al. (1991). "Detection of specific polymerase chain reaction product by utilizing the $5' \rightarrow 3'$ exonuclease activity of Thermus aquaticus DNA polymerase." *Proc Natl Acad Sci*USA 88(16): 7276-7280.

Isacsson, J., H. Cao, et al. (2000). "Rapid and specific detection of PCR products using light-up probes." *Molecular and cellular probes*14(5): 321-328.

Jansen, K., B. Norde´n, et al. (1993). "Sequence dependence of 40,6-diamidino-2-phenylindole (DAPI)-DNA interactions." *J Am Chem Soc*115: 10527-10530.

Klein, D. (2002). "Quantification using real-time PCR technology: applications and limitations." *Trends Mol Med*8(6): 257-260.

Konigshoff, M., J. Wilhelm, et al. (2003). "HER-2/neu gene copy number quantified by real-time PCR: comparison of gene amplification, heterozygosity, and immunohistochemical status in breast cancer tissue." *Clin Chem*49(2): 219-229.

Krauter, J., G. Heil, et al. (2001). "The AML1/MTG8 Fusion Transcript in t (8; 21) Positive AML and its Implication for the Detection of Minimal Residual Disease; Malignancy." *Hematology (Amsterdam, Netherlands)*5(5): 369-381.

Krauter, J., W. Hoellge, et al. (2001). "Detection and quantification of CBFB/MYH11 fusion transcripts in patients with inv(16)-positive acute myeloblastic leukemia by real-time RT-PCR." *Genes Chromosomes Cancer*30(4): 342-348.

Kubista, M. (2004). "Nucleic acid-based technologies: application amplified." *Pharmacogenomics*5(6): 767-773.

Kubista, M., J. M. Andrade, et al. (2006). "The real-time polymerase chain reaction." *Molecular Aspects of Medicine*27 95-125.

Kurata, S., T. Kanagawa, et al. (2001). "Fluorescent quenching-based quantitative detection of specific DNA/RNA using a BODIPY® FL-labeled probe or primer." *Nucleic Acids Research*29(6): e34.

Kutyavin, I. V., I. A. Afonina, et al. (2000). "3´-Minor groove binder-DNA probes increase sequence specificity at PCR extension temperatures." *Nucleic Acids Res*28(2): 655-661.

Lay, M. J. and C. T. Wittwer (1997). "Real-time fluorescence genotyping of factor V Leiden during rapid-cycle PCR." *Clin Chem*43(12): 2262-2267.

Le Pecq, J. B. and C. Paoletti (1966). "A new fluorometric method for RNA and DNA determination." *Anal Biochem*17: 100-107.

Lehmann, U., S. Glöckner, et al. (2000). "Detection of gene amplification in archival breast cancer specimens by laser-assisted microdissection and quantitative real-time polymerase chain reaction." *Am J Pathol*156(6): 1855-1864.

Li, Qingge, et al. (2002). "A new class of homogeneous nucleic acid probes based on specific displacement hybridization." *Nucleic Acids Res*30: e5.

Lin, M. H., C. H. Tseng, et al. (2001). "Real-time PCR for rapid genotyping of angiotensin-converting enzyme insertion/deletion polymorphism1." *Clinical biochemistry*34(8): 661-666.

Lind, K., A. Stahlberg, et al. (2006). "Combining sequence-specific probes and DNA binding dyes in real-time PCR for specific nucleic acid quantification and melting curve analysis." *Biotechniques*40(3): 315-319.

Liss, B. (2002). "Improved quantitative real-time RT-PCR for expression profiling of individual cells." *Nucleic Acids Res*30(17): e89.

Liu, W. and D. A. Saint (2002). "A new quantitative method of real time reverse transcription polymerase chain reaction assay based on simulation of polymerase chain reaction kinetics." *Analytical Biochemistry*302(1): 52-59.

Liu, W. and D. A. Saint (2002). "Validation of a quantitative method for real time PCR kinetics." *Biochem Biophys Res Commun*294(2): 347-353.

Livak, K. J. and T. D. Schmittgen (2001). "Analysis of relative gene expression data using real-time quantitative PCR and the 2(-Delta Delta C(T)) Method." *Methods*25(4): 402-408.

Locatelli, G., V. Urso, et al. (2000). "Quantitative analysis of GMO food contaminations using real time PCR." *Ital J Biochem*49(3-4): 61-63.

Luu-The, V., N. Paquet, et al. (2005). "Improved real-time RT-PCR method for high-throughput measurements using second derivative calculation and double correction." *Biotechniques*38(2): 287-293.

Lyon, E., A. Millson, et al. (2001). "Quantification of HER2/neu gene amplification by competitive PCR using fluorescent melting curve analysis." *Clinical chemistry*47(5): 844-851.

Maas, F., N. Schaap, et al. (2003). "Quantification of donor and recipient hemopoietic cells by real-time PCR of single nucleotide polymorphisms." *Leukemia*17(3): 621-629.

Mackay, I. M. (2004). "Real-time PCR in the microbiology laboratory." *Clin Microbiol Infect*10(3): 190-212.

Mackay, I. M., K. E. Arden, et al. (2002). "Real-time PCR in virology." *Nucleic Acids Res*30(6): 1292-1305.

Mackya, I. M. (2004). "Real-time PCR in the microbiology laboratory." *Clin Microbiol Infect*10: 190-212.

Marcucci, G., K. Livak, et al. (1998). "Detection of minimal residual disease in patients with AML1/ETO-associated acute myeloid leukemia using a novel quantitative reverse transcription polymerase chain reaction assay." *Leukemia: official journal of the Leukemia Society of America, Leukemia Research Fund, UK*12(9): 1482-1489.

Mattarucchi, E., M. Marsoni, et al. (2005). "Different real time PCR approaches for the fine quantification of SNP's alleles in DNA pools: assays development, characterization and pre-validation." *J Biochem Mol Biol*38(5): 555-562.

Mehmet, D., F. Ahmed, et al. (2001). "Quantification of the common deletion in human testicular mitochondrial DNA by competitive PCR assay using a chimaeric competitor." *Molecular human reproduction*7(3): 301-306.

Mhlanga, M. M. and L. Malmberg (2001). "Using molecular beacons to detect single-nucleotide polymorphisms with real-time PCR." *Methods*25(4): 463-471.

Miyazato, A., S. Ueno, et al. (2001). "Identification of myelodysplastic syndrome–specific genes by DNA microarray analysis with purified hematopoietic stem cell fraction." *Blood*98(2): 422-427.

Miyoshi, Y., A. Ando, et al. (2002). "Prediction of response to docetaxel by CYP3A4 mRNA expression in breast cancer tissues." *International Journal of Cancer*97(1): 129-132.

Muller, P. Y., H. Janovjak, et al. (2002). "Processing of gene expression data generated by quantitative real-time RT-PCR." *Biotechniques*32(6): 1372-1379.

Mullis, K. B. (1990). "The unusual origin of the polymerase chain reaction." *Sci Am*262: 56-61.

Mullis, K. B. and F. A. Faloona (1987). "Specific synthesis of DNA in vitro via a polymerase-catalyzed chain reaction." *Methods Enzymol*155: 335-350.

Myi, Q. "single-color real-time PCR instruction manual Biorad Company."

Nazarenko, I., B. Lowe, et al. (2002). "Multiplex quantitative PCR using self-quenched primers labelled with a single fluorophore." *Nucleic Acids Res*30(9): e37.

Nazarenko, I. A., S. Bhatnagar, et al. (1997). "A closed tube format for amplification and detection of DNA based on energy transfer." *Nucleic Acids Research*25(12): 2516-2521.

Nigro, J. M., M. A. Takahashi, et al. (2001). "Detection of 1p and 19q loss in oligodendroglioma by quantitative microsatellite analysis, a real-time quantitative polymerase chain reaction assay." *The American journal of pathology*158(4): 1253-1262.

Norton, D. M. (2002). "Polymerase chain reaction-based methods for detection of Listeria monocytogenes: toward real-time screening for food and environmental samples." *Journal of AOAC International*85(2): 505-515.

Nygren, J., N. Svanvik, et al. (1998). "The interaction between the fluorescent dye thiazole orange and DNA." *Biopolymers*46: 39-51.

Olson, M., L. Hood, et al. (1989). "A common language for physical mapping of the human genome." *Science*245: 1434-1435.

Palladino, S., I. D. Kay, et al. (2003). "Real-time PCR for the rapid detection of vanA and vanB genes." *Diagn Microbiol Infect Dis*45(1): 81-84.

Pfaffl, M. W. (2001). "A new mathematical model for relative quantification in real-time RT-PCR." *Nucleic Acids Res*29(9): e45.

Pfaffl, M. W., A. Tichopad, et al. (2004). "Determination of stable housekeeping genes, differentially regulated target genes and sample integrity: BestKeeper–Excel-based tool using pair-wise correlations." *Biotechnology letters*26(6): 509-515.

Poola, I. (2003). "Molecular assay to generate expression profile of eight estrogen receptor alpha isoform mRNA copy numbers in picogram amounts of total RNA from breast cancer tissues." *Analytical Biochemistry*314(2): 217-226.

Prieto-Alamo, M. J., J. M. Cabrera-Luque, et al. (2003). "Absolute quantitation of normal and ROS-induced patterns of gene expression: an in vivo real-time PCR study in mice." *Gene Expr*11(1): 23-34.

Randen, I., K. Sørensen, et al. (2003). "Rapid and reliable genotyping of human platelet antigen (HPA)-1,-2,-3,-4, and-5 a/b and Gov a/b by melting curve analysis." *Transfusion*43(4): 445-450.

Reimer, T., D. Koczan, et al. (2002). "Tumour Fas ligand: Fas ratio greater than 1 is an independent marker of relative resistance to tamoxifen therapy in hormone receptor positive breast cancer." *Breast Cancer Research*4(5): R9.

Reischl, U., C. T. Wittwer, et al. (2002). "Rapid Cycle Real-time PCR: Methods and Applications; Microbiology and Food Analysis." *New York: Springer-Verlag.*

Rickman, D. S., M. P. Bobek, et al. (2001). "Distinctive molecular profiles of high-grade and low-grade gliomas based on oligonucleotide microarray analysis." *Cancer research*61(18): 6885-6891.

Ririe, K. M., R. P. Rasmussen, et al. (1997). "Product differentiation by analysis of DNA melting curves during the polymerase chain reaction." *Anal Biochem*245(2): 154-160.

Ruiz-Ponte, C., L. Loidi, et al. (2000). "Rapid real-time fluorescent PCR gene dosage test for the diagnosis of DNA duplications and deletions." *Clin Chem*46(10): 1574-1582.

Saiki, R. K., S. Scharf, et al. (1985). "Enzymatic amplification of beta-globin genomic sequences and restriction site analysis for diagnosis of sickle cell anemia." *Science*230: 1350-1354.

Sails, A. D., A. J. Fox, et al. (2003). "A real-time PCR assay for the detection of Campylobacter jejuni in foods after enrichment culture." *Appl Environ Microbiol*69(3): 1383-1390.

Sasvari-Szekely, M., A. Gerstner, et al. (2000). "Rapid genotyping of factor V Leiden mutation using single-tube bidirectional allele-specific amplification and automated ultrathin-layer agarose gel electrophoresis." *Electrophoresis*21(4): 816-821.

Shively, L., L. Chang, et al. (2003). "Real-time PCR assay for quantitative mismatch detection." *Biotechniques*34(3): 498-502, 504.

Sjöback, R., J. Nygren, et al. (1995). "Absorption and fluorescence properties of fluorescein Spetrochim." *Acta Part*A 51: L7-L21.

Soheili, Z. and S. Samiei (2005). "Real Time PCR: Principles and Application." *Hepatitis*5(3): 83-87.

Solinas, A., L. J. Brown, et al. (2001). "Duplex Scorpion primers in SNP analysis and FRET applications." *Nucleic Acids Research*29(20): e96-e96.

Souaze, F., A. Ntodou-Thome, et al. (1996). "Quantitative RT-PCR: limits and accuracy." *Biotechniques*21(2): 280-285.

Stordeur, P., L. F. Poulin, et al. (2002). "Cytokine mRNA quantification by real-time PCR." *J Immunol Methods*259(1-2): 55-64.

Svanvik, N., A. Stahlberg, et al. (2000). "Detection of PCR products in real time using light-up probes." *Anal Biochem*287(1): 179-182.

Svanvik, N., G. Westman, et al. (2000). "Light-up probes: thiazole orange-conjugated peptide nucleic acid for detection of target nucleic acid in homogenous solution." *Anal Biochem*281: 26-35.

Thelwell, N., S. Millington, et al. (2000). "Mode of action and application of Scorpion primers to mutation detection." *Nucleic Acids Research*28(19): 3752-3761.

Tyagi, S., D. P. Bratu, et al. (1998). "Multicolor molecular beacons for allele discrimination." *Nat Biotechnol*16(1): 49-53.

Tyagi, S. and F. R. Kramer (1996). "Molecular Beacons: probes that fluorescence upon hybridization." *Nat Biotechnol*14(3): 303-308.

Uehara, H., G. Nardone, et al. (1999). "Detection of telomerase activity utilizing energy transfer primers: comparison with gel- and ELISA-based detection." *Biotechniques*26(3): 552-558.

Ulvik, A. and P. M. Ueland (2001). "Single nucleotide polymorphism (SNP) genotyping in unprocessed whole blood and serum by real-time PCR: application to SNPs affecting homocysteine and folate metabolism." *Clin Chem*47(11): 2050-2053.

Valasek, M. A. and J. J. Repa (2005). "The power of real-time PCR." *Adv Physiol Educ*29(3): 151-159.

Verhagen, O. J., M. J. Willemse, et al. (2000). "Application of germline IGH probes in real-time quantitative PCR for the detection of minimal residual disease in acute lymphoblastic leukemia." *Leukemia*14(8): 1426-1435.

von Wurmb-Schwark, N., R. Higuchi, et al. (2002). "Quantification of human mitochondrial DNA in a real time PCR." *Forensic Sci Int*126(1): 34-39.

Whitcombe, D., J. Brownie, et al. (1998). "A homogeneous fluorescence assay for PCR amplicons: its application to real-time, single-tube genotyping." *Clin Chem*44(5): 918-923.

Whitcombe, D., J. Theaker, et al. (1999). "Detection of PCR products using self-probing amplicons and fluorescence." *Nat Biotechnol*17(8): 804-807.

Wilhelm, J. and A. Pingoud (2003). "Real-time polymerase chain reaction." *Chembiochem*4(11): 1120-1128.

Wilhelm, J., A. Pingoud, et al. (2003). "Real-time PCR-based method for the estimation of genome sizes." *Nucleic Acids Res*31(10): e56.

Wilhelm, J., A. Pingoud, et al. (2003). "SoFAR: software for fully automatic evaluation of real-time PCR data." *Biotechniques*34(2): 324-332.

Wilhelm, J., H. Reuter, et al. (2002). "Detection and quantification of insertion/deletion variations by allele-specific real-time PCR: application for genotyping and chimerism analysis." *Biol Chem*383(9): 1423-1433.

Wilson, I. G. (1997). "Inhibition and facilitation of nucleic acid amplification." *Appl Environ Microbiol*63: 3741-3751.

Wilson, R. and M. K. Johansson (2003). "Photoluminescence and electrochemiluminescence of a Ru(II) (bpy)3- quencher dual-labeled oligonucleotide probe." *Chem Commun*21: 2710-2711.

Wittwer, C. T., M. G. Herrmann, et al. (2001). "Real-time multiplex PCR assays." *Methods*25(4): 430-442.

Wittwer, C. T., M. G. Herrmann, et al. (1997). "Continuous fluorescence monitoring of rapid cycle DNA amplification." *Biotechniques*22: 130-138.

Wong, M. L. and J. F. Medrano (2005). "Real-time PCR for mRNA quantitation." *Biotechniques*39(1): 75-85.

Worm, J., A. Aggerholm, et al. (2001). "In-tube DNA methylation profiling by fluorescence melting curve analysis." *Clinical chemistry*47(7): 1183 -1189.

Ye, J., E. J. Parra, et al. (2002). "Melting Curve SNP (McSNP) Genotying: a Useful Approach for Diallelic Genotyping in Forensic Science." *Journal of forensic sciences*47(3): 593-600.

Zipper, H., H. Brunner, et al. (2004). "Investigations on DNA intercalation and surface binding by SYBR Green I, its structure determination and methodological implications." *Nucleic Acids Res*32(12): e103.

# 8

# PCR Advances Towards the Identification of Individual and Mixed Populations of Biotechnology Microbes

P. S. Shwed

*Biotechnology Laboratory, Environmental Health Sciences and Research Bureau,*
*Environmental and Radiation Health Sciences Directorate,*
*Healthy Environments and Consumer Safety Branch, Health Canada,*
*Canada*

## 1. Introduction

Public health and safety, diagnostics and surveillance are aided by knowledge of the identity and genetic content of biotechnology microbes and their close relatives. Both types of information allow recognition and prediction of virulence and pathogenicity of microbes. PCR has played an important role in enabling the identification of micro-organisms and the distinction of pathogenic from non-pathogenic species, since the technical descriptions in the mid-1980s (Mullis et al., 1986; Mullis and Faloona, 1987). This DNA amplification technology allows the generation of large template quantity, a pre-requisite for cloning and for dideoxy DNA "Sanger" sequencing (Sanger et al., 1977). As such, PCR has been integral in first generation and phylogenetic marker sequencing projects (Bottger, 1989).

During the last decade, PCR has remained a cornerstone in microbial genetic characterization. Marker sequencing remains a component of the polyphasic characterization of microbial genomes in which genetic, morphological and biochemical data are reconciled. At the same time, great progress has been made in single cell microbial genetics and PCR miniaturization has been implemented in second generation sequencing platforms (Metzker, 2010). Collectively, these developments have resulted in increased numbers of whole genome sequences from individual microbes of "unculturable" microorganisms and outbreak strains such as Shiga toxin–producing *E. coli* strain O104:H4 detected in Europe during 2011 (Mellmann et al., 2011). High throughput sequencing has allowed for insights into natural and human environments and their mixed bacterial populations (Hamady and Knight, 2009; Mardis, 2011; Sapkota et al., 2010).

This chapter serves to highlight PCR advances that have enabled microbial identification during the last decade. At the level of single species, identifications can involve phylogenetic marker sequencing, or whole genome sequencing from individual cells or cultures. Mixed microbial populations, may be sorted, individually identified by sequencing or collectively sequenced using high throughput platforms. The potential and challenges of these new platforms, as well as their applications towards novel microbial strains that will be produced by synthetic biology approaches, will be discussed.

## 2. Current challenges in microbial identification

Collectively, microbes occupy a vast range of ecological niches and feature intrinsic diverse metabolic potential. Microbial biotechnology has enabled the screening and enhancement of strains for commercial applications such as: preservation and harvest of natural resources (bio-pesticides and bio-mining of metals), environmental remediation (improved soil/air/water quality) and applications for sustainable development. Often, biotechnology microbes are used as single species, while other commercial products involve mixtures of a few or many different species and strains.

Bacterial strains, that feature desirable phenotypic traits, have been traditionally isolated by high-throughput screening, or strains have been improved by random mutagenesis and screening. Currently, consensus identification and classification of bacterial strains is carried out by a polyphasic approach. Phenotypic data (biochemical tests, fatty acid composition), genotypic data and phylogenetic information, that includes genetic information, derived from PCR amplification of marker genes, are reconciled (Vandamme et al., 1996).

Discrimination of beneficial and harmful species is challenging in a number of genera that contain closely related species: *Burkholderia, Bacillus, Acinetobacter and Pseudomonas*. For example, in the *Burkholderia* genus, *B. cepacia* is a non-pathogenic soil bacterium that is being developed for the application of phytoremediation (Barac et al., 2004) and clinically, *B. cepacia* bacteria have been associated with infections and cystic fibrosis, as reviewed in (Coenye and Vandamme, 2003). Another prime example concerns the *Bacillus cereus sensu lato* family of bacteria. This group comprises the *Bacillus cereus* species *sensu stricto, B. anthracis*, strains and subspecies of *B. thuringiensis, B. mycoides*, and *B. weihenstephanensis*. Most *Bacillus cereus* organisms are common soil bacteria that are pathogenic to insects and invertebrates. Some species may cause contamination problems in the dairy industry and paper mills and may also be a causative agent of food poisoning. Select strains of *Bacillus anthracis* and a few *Bacillus cereus sensu stricto* strains are the only ones reported to cause fatal pulmonary or intestinal infections (Dixon et al., 2000).

Bacterial mixtures, that have been created or isolated in order to carry out a function, pose technical challenges to polyphasic characterization. Phenotypic data is typically derived for individual microbial isolates. However, current culture techniques cannot support a substantial fraction of microbial species (Handelsman, 2004) and there is risk of bias towards culturable species. In these cases, molecular methods that directly acquire genetic information may remove this hindrance.

### 2.1 PCR towards microbial identification

### 2.1.1 Phylogenetic marker amplification and analysis

Of all global markers, small subunit ribosomal RNA (16S rRNA) encoding genes are the best characterized genes for microbial systematics. The 16S rRNA gene is ubiquitous, highly conserved, but possesses enough variability to allow taxa specific discrimination. The gene is composed of nine hypervariable regions separated by conserved regions and sequences are available for numerous organisms via public databases such as NCBI, the Ribosomal Database project (Cole et al., 2009) and Greengenes (Desantis et al., 2006). The nine different variable 16S rRNA regions are flanked by conserved nucleotide stretches in bacteria (Neefs et al., 1993) and these could be used as targets for PCR primers with near-universal specificity.

It was during the mid-1980's that PCR first enabled molecular microbial ecology studies involving the 16S rRNA gene. Pace and colleagues first amplified 16S rRNA from bulk nucleic acid extractions using nearly "universal" primers, in order to sequence, classify and compare these to phylogenetic trees (Pace, 1997) (Lane et al., 1985) (Woese, 1987). At this time it was observed that not all environmental microorganisms were capable of colony formation and that by sequencing cloned ribosomal DNA, new microbial species could be revealed (Stahl et al., 1984),(Stahl et al., 1985).

Over the last few decades, a large number of primer sequences have been designed for amplification and sequencing of 16S RNA genes, as reviewed in (Baker et al., 2003). There are a number of databases available for the primer sequences. Some of these primers have been designed as taxa specific, whereas others have been designed to amplify all prokaryotic rRNA genes and are referred to as "universal".

16S rRNA sequences may offer limited taxonomic resolution, particularly for genera that feature close phylogenetic relationships. *B. cepacia* complex reference strains feature high similarity values (above 98%) which reflects a close phylogenetic relationship (Coenye and Vandamme, 2003). Also, up to 2% intraspecies diversity has been observed in *B. cepacia* rRNA sequences and they cannot be identified at the species level by simple comparison of 16S rRNA sequences. Similarly, for the *B. cereus* group, there is also insufficient divergence in 16S rRNA to allow for resolution of strains and species (Bavykin et al., 2004). In these cases, other global markers have been explored for strain discrimination such as the genes that encode: RNA polymerase subunits, DNA gyrases, heat shock and recA proteins and hisA. The strong functional and structural constraints for these gene products, limits the number of mutations that can occur in the genes and renders them useful as markers for relatedness.

Identification of distinct strains of a prokaryotic species can take place by multi-locus sequence typing, in which sequence mismatches in a small number of house keeping genes are analyzed (as reviewed in (Maiden, 2006)). In the case of prokaryotic identification of closely related species, a similar strategy designated multi-locus sequence analysis has been used for several studies and involves a two step process: rRNA sequencing in order to assign an unknown strain to a group (either genus or family), that in turn defines the particular genes and primers to be used for analysis. This two-tiered approach has allowed discrimination of *Burkholderia* strains and those of the *Bacillus cereus* group (reviewed in (Gevers et al., 2005)).

### 2.1.2 PCR as a component of genomic methods

During the last decade, various applications of DNA microarrays have been used to assess the risk of a particular microbe by enabling detection and/or identification at the species, subspecies or strain level, or presence of virulence genes (reviewed in (Shwed et al., 2007)). However, DNA amplification is rarely a technical component of these studies. However, as will be described in section 3.1, novel PCR amplification strategies are a component of the workflow for high throughput sequencing platforms.

## 3. Miniaturization of PCR

Arguably, the major PCR advancement of the last decade has been the development of miniaturized and parallelized platforms. Whereas previously PCR reactions were typically carried out at the microlitre scale, new configurations have enabled femtolitre scale reactions. In turn, higher throughput and cost efficiencies have been achieved.

One miniaturization has been achieved by reaction entrapment in thermodynamically stable "water in oil" nanoreactor microemulsion systems, such as reverse micelles, as described for enzymatic reactions (Klyachko and Levashov, 2003). These emulsions are easily prepared and stable under a wide variety of temperatures, pH and salt concentrations. The smallest droplets rival the scale of bacteria with diameters of less than one micrometre with volumes in the femtolitre scale.

Emulsion PCR was first reported for the directed evolution of heat-stable, heparin insensitive variants of *Taq* DNA polymerase (Ghadessy et al., 2001). The concept of emulsion PCR was to disperse template DNA into a water in oil emulsion such that most droplets contain a single template and only a few droplets contain more than one template. Amplification was carried out within the drops by PCR, so that each droplet generated an amplified number of clonal copies.

### 3.1 Convergence of miniaturized PCR with other technologies

During the last decade, advancements have been made in the engineering of microfluidic scale devices that integrate multiple analytical steps into "laboratory on chip" systems (as reviewed in (Liu and Mathies, 2009)). These devices allow the generation and manipulation of aqueous microdroplets at high rates and with high fidelity manipulation in microfluidic channels. PCR-based genetic analysis and sequencing can now be carried out at the picolitre to nanolitre volume scale, with the advantages of decreased thermal cycling times and reagent consumption along with increased throughput.

Microfluidic droplet PCR has been reported to allow 1.5 million parallel amplifications for target enrichment of loci in the human genome (Tewhey et al., 2009). In this instance, microfluidic chips were designed to merge 20 picolitre droplets that contain about 3 picograms of biotinylated fragments of template DNA (2-4 kb) with droplets that contain a pair of PCR primers that amplify specific sequences. This platform allowed a yield of more than one million merged droplets that are subjected to PCR. At the end of the amplification reaction, the emulsion is broken. After centrifugation, the aqueous phase, that contains the PCR products from all the droplets, is subjected to a second generation sequencing strategy.

During the last decade, several commercial second-generation sequencing platforms have been developed and these feature cyclic array sequencing strategies, involving new variations of PCR. In both cases, amplification of densely arrayed amplicons is achieved, in order to serve as features for *in situ* sequencing and imaging-based sequence by synthesis data collection (more detailed descriptions of second generation sequencing platforms are reviewed in (Shendure et al., 2011)). Common to all strategies, the first step is the *in vitro* generation of a shot gun genomic library, by the random fragmentation of DNA and the ligation of universal adaptor sequences. Afterwards, *in vitro* clonal amplification is carried out by one of two principal types of PCR, which generate template for sequencing. Table 1 shows how various commercial platforms use PCR to derive features that are sequenced.

Emulsion PCR is carried out as described above (section 3.0 and shown in Fig.1 A, B), with the exception that paramagnetic beads that are bound to one of the PCR primers on their surface, are used (Dressman et al., 2003). These beads allow the solid-phase capture of clonally amplified PCR amplicons from each emulsion PCR compartment. For some commercial pyrosequencing platforms, beads are then deposited on microfabricated arrays of picolitre scale wells to allow immobilization and *in situ* pyrosequencing.

Bridge PCR (Adessi et al., 2000; Fedurco et al., 2006) involves the use of spatially distributed oligonucleotides that are covalently attached to a support (shown in Fig. 1 C,D). A DNA library is hybridized as single stranded DNA to the support. Immobilized copies of the library are synthesized by extension from the immobilized primers. After denaturation, the template copies are able to loop and hybridize to an adjacent oligonucleotide on the support. Additional copies of the template are synthesized and the process is repeated on each template so that clonal clusters, each with about 2000 molecules are generated.

| Instrument | PCR type | Sequence Method | Reference |
|---|---|---|---|
| 454 | Emulsion | Pyrosequencing | (Margulies et al., 2005) |
| Illumina | Bridge | Polymerase | (Fedurco et. al. 2006) |
| SOLiD | Emulsion | Ligase | (Shendure et al., 2005) |

Table 1. PCR clonal amplification by second-generation sequencing instrument

## 4. Emerging PCR applications

### 4.1 Second generation sequencing from microbial mixtures

In recent years, complex microbial communities, such as those of the human gut intestinal tract, or those associated with biofilm infections, have been analyzed by second generation sequencing of shot gun libraries derived from either metagenomic DNA, or PCR amplified variable 16S regions amplified from metagenomic DNA prepared from a microbial mixture (Arumugam et al., 2011; Dowd et al., 2008).

Second generation platforms allow economies of scale in sequencing. PCR amplified products can be characterized without cloning, which saves time and costs. Also, the estimated costs per megabase of derived sequence are lower for the new platforms compared to first generation sequencing (Shendure et. al. 2011). Lastly, multiplexed runs, derived from 16S rRNA coding sequences from several communities, are feasible by using unique sequence barcodes during amplification (Hamady et al., 2008).

It has been proposed that sequencing of individual variable regions is sufficient for taxonomic differentiation of bacterial mixtures (Liu et al., 2007). The sequence read lengths of second generation platforms are generally short, but several new models have shown greater read lengths (Liu et al., 2008). On the other hand, direct sequencing of metagenomic DNA has been proposed to be less biased than that of PCR amplified DNA, due to lack of 16S primer bias (von Mering et al., 2007).

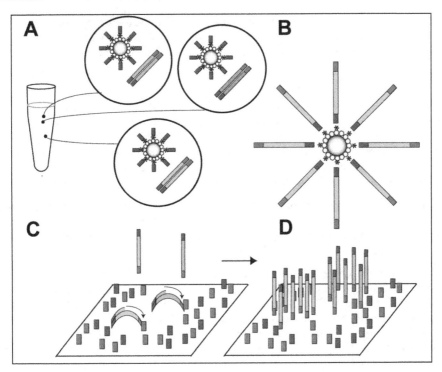

Fig. 1. PCR advancements towards second-generation sequencing

**Panels A.B: Emulsion PCR**

Panel A) A shot-gun DNA library is ligated to adaptors (blue and red bars), diluted, and PCR amplified in a water in oil emulsion, within aqueous microdroplets. The droplets contain streptavidin coated beads that carry one of the biotinylated PCR primers tethered to beads. Panel B) Where DNA is amplified in the presence of a bead, several thousand copies of the template will be captured.

**Panels C,D: Bridge PCR**

Panel C) A shot-gun DNA library is ligated to adaptors, made single stranded and hybridized to PCR primers that are immobilized with flexible linkers on a substrate. Bridge amplification occurs when primer extension occurs from immediately adjacent primers. Panel D) Immobilized clusters of about one thousand amplicons are formed after successive cycles of extension and denaturation.

The critical analytical step of taxonomic analyses of microbial diversity analysis is known as binning, where the sequences from a mixture of organisms are assigned phylogenetic groups. However, the outcome of binning results may range from kingdom level to genus level assignment, depending on the quality of data and the read length of data (Yang et al., 2010). One of the binning strategies in use is based on classification of DNA fragments based on sequence homology, using publically available reference databases such as Basic Local Alignment Search Tool (Huson et al., 2007; Meyer et al., 2008). The second strategy involves similarity to protein families and domains, such as in the phylogenetic algorithm CARMA (Krause et al., 2008).

Collectively, these identification approaches are limited by the use of reference databases of known species and genes from readily cultivated microbes. As a consequence, species within a microbial community that lack a reference sequence will remain unidentified.

## 4.2 PCR analysis of single cells

The analysis of complex mixtures of environmental bacteria will benefit from microfluidic digital PCR analysis that involves single cell sorting from mixtures of bacteria. Single bacterial cells can be isolated by various technologies, including: optical tweezers, micromanipulation, FACS, serial dilutions, or laser capture microdissection. In turn, experimentation that involves retrieving "needles in a haystack", such as searches for microbes featuring particular genes are facilitated by microfluidics technologies (Baker, 2010).

Characterization of environmental bacteria of the 1 microlitre volume termite hindgut model, exemplify the potential of cell sorting and PCR. This microenvironment contains about $10^6$-$10^8$ microbial cells, comprised of unculturable species not detected in other environments (reviewed in (Hongoh, 2010)).

Otteson et al. (Ottesen et al., 2006), applied a microfluidic digital PCR characterization approach for the termite bacteria. In this study, individual cells were partitioned in a microfluidic array panel and served as templates for the simultaneous amplification of both rRNA and metabolic genes of interest. The digital PCR aspect involved ensuring that the partitioning was into reactions that contained an average of one template (bacterial cell) or less (Sykes et al., 1992). Retrieved PCR products from individual chambers allowed sequence analysis of both genes by standard methods and allowed the determination of new bacterial species that contribute to metabolism. More recently, microfluidic digital PCR was used to associate particular viruses that infect the bacteria of the termite gut, without culturing either the viruses or the hosts (Tadmor et al., 2011). Here, amplification of both rRNA gene and a viral marker gene was carried out from a PCR array panel containing individual microbes.

## 4.3 Whole genome sequencing from individual cells

Genomic sequences provide the most absolute indication of genetic variation and virulence potential for a bacterial strain. The documentation of the complete nucleic acid sequences of high priority beneficial and detrimental microorganisms in public databases are efforts that can greatly aid the identification of unknown strains. In studies involving closely related bacterial strains, shotgun library sequences can be assembled by mapping the reads to a reference genome.

Direct single bacterial cell genome sequencing can be carried out by multiple displacement amplification, using individually lysed bacteria and the few femtograms of DNA present in bacterial cells in order to generate template for shotgun sequencing. This reaction involves the use of $\phi$29 DNA polymerase and random primers to amplify DNA templates under isothermal conditions (Dean et al., 2001).

Genomic sequencing from individual uncultured bacterial cells was first shown by Raghunathan et al., using *E. coli* cells that had been isolated by flow cytometry (Raghunathan et al., 2005). This report illustrated contamination as a technical challenge when working with individual microbial cells. The reaction involves random primers in order to initiate polymerization and this can result in amplification of contaminating DNA. In the case of poorly characterized or novel biotechnology microbes, the non-target DNA could confound conclusions about the target organism. In addition, there are biases introduced by multiple displacement amplification, particularly with the use of small input quantity of DNA. Segments of the chromosomes have been observed to be preferentially amplified. As well, chimeric rearrangements of DNA result from the linking of non-contiguous chromosomal regions (Zhang et al., 2006).

Despite these challenges, there have been recent reports that are more encouraging about the acquisition of finished genomic sequence derived from a single bacterial cell (Woyke et al., 2010). Multiple displacement amplification artifacts have been overcome with new computational algorithms, that can compensate for amplification bias and chimeric sequences, using short sequence reads (Chitsaz et al., 2011).

## 5. Conclusions and future challenges

The safe use of biotechnology microbes for public health and in the environment requires knowledge of the identity and genetic potential of these organisms. In the first decade of the 21st century, amongst the genetic tools available for genetic characterization, PCR remains a cornerstone. Advances in miniaturization and parallelism of PCR have enhanced throughput and enabled second generation sequencing platforms. These technological advancements have been linked to progress in single cell microbial genomics, whole genome sequencing and the characterization of microbial mixtures. Collectively, these developments have direct implications for the safety assessments that are carried out by industry and governments.

These recent technological advances will allow new human and environmental surveys. As an example, movements of genes amongst microbes by horizontal gene transfer mechanisms may be tracked. Environmental surveys of the movements of particular nucleotide sequences are now possible by metagenomic methods. Culture-independent methodology for genetic analysis will allow greater throughput. However, at present, computational hurdles remain for the wide-spread implementation of such technology.

Miniaturization has been a hallmark of progress in electronics and computing. By this measure, PCR miniaturization that has taken place to date is of relatively low order. At the same time, the complexity of biotechnology microbes developed for commercial applications is increasing. The advances in PCR and genomic technologies must be considered in parallel with the technical advancements that have been made towards the *de novo* construction of synthetic microbes. High throughput, high efficiency microfluidic devices can enable the encapsulation of novel genetic material in abiotic chassis (Szita et al., 2010). PCR and sequencing advancements will remain important for microbial genetic characterization.

## 6. Acknowledgements

Drs. Guillaume Pelletier and Azam Tayabali are thanked for constructive criticism of the manuscript. Open access charges are supported by the Canadian Regulatory System for Biotechnology Fund.

## 7. References

Adessi C, Matton G, Ayala G, Turcatti G, Mermod JJ, Mayer P, and Kawashima E. 2000. Solid phase DNA amplification: characterisation of primer attachment and amplification mechanisms. *Nucleic Acids Res.* 28 (20): E87.

Arumugam M, Raes J, Pelletier E, Le PD, Yamada T, Mende DR, Fernandes GR, Tap J, Bruls T, Batto JM, Bertalan M, Borruel N, Casellas F, Fernandez L, Gautier L, Hansen T, Hattori M, Hayashi T, Kleerebezem M, Kurokawa K, Leclerc M, Levenez F, Manichanh C, Nielsen HB, Nielsen T, Pons N, Poulain J, Qin J, Sicheritz-Ponten T, Tims S, Torrents D, Ugarte E, Zoetendal EG, Wang J, Guarner F, Pedersen O, de Vos WM, Brunak S, Dore J, Antolin M, Artiguenave F, Blottiere HM, Almeida M, Brechot C, Cara C, Chervaux C, Cultrone A, Delorme C, Denariaz G, Dervyn R, Foerstner KU, Friss C, van de GM, Guedon E, Haimet F, Huber W, van Hylckama-Vlieg J, Jamet A, Juste C, Kaci G, Knol J, Lakhdari O, Layec S, Le RK, Maguin E, Merieux A, Melo MR, M'rini C, Muller J, Oozeer R, Parkhill J, Renault P, Rescigno M, Sanchez N, Sunagawa S, Torrejon A, Turner K, Vandemeulebrouck G, Varela E, Winogradsky Y, Zeller G, Weissenbach J, Ehrlich SD, and Bork P. 2011. Enterotypes of the human gut microbiome. *Nature* 473 (7346): 174-180.

Baker GC, Smith JJ, and Cowan DA. 2003. Review and re-analysis of domain-specific 16S primers. *J Microbiol. Methods* 55 (3): 541-555.

Baker M. 2010. Clever PCR: more genotyping, smaller volumes. *Nat Meth* 7 (5): 351-356.

Barac T, Taghavi S, Borremans B, Provoost A, Oeyen L, Colpaert JV, Vangronsveld J, and van der LD. 2004. Engineered endophytic bacteria improve phytoremediation of water-soluble, volatile, organic pollutants. *Nat Biotechnol.* 22 (5): 583-588.

Bavykin SG, Lysov YP, Zakhariev V, Kelly JJ, Jackman J, Stahl DA, and Cherni A. 2004. Use of 16S rRNA, 23S rRNA, and *gyrB* gene sequence analysis to determine phylogenetic relationships of *Bacillus cereus* group microorganisms. *J Clin. Microbiol.* 42 (8): 3711-3730.

Bottger EC. 1989. Rapid determination of bacterial ribosomal RNA sequences by direct sequencing of enzymatically amplified DNA. *FEMS Microbiol. Lett.* 53 (1-2): 171-176.

Chitsaz H, Yee-Greenbaum JL, Tesler G, Lombardo MJ, Dupont CL, Badger JH, Novotny M, Rusch DB, Fraser LJ, Gormley NA, Schulz-Trieglaff O, Smith GP, Evers DJ, Pevzner PA, and Lasken RS. 2011. Efficient *de novo* assembly of single-cell bacterial genomes from short-read data sets. *Nat. Biotechnol.* 29 (10): 915-921

Coenye T, and Vandamme P. 2003. Diversity and significance of Burkholderia species occupying diverse ecological niches. *Environ. Microbiol.* 5 (9): 719-729.

Cole JR, Wang Q, Cardenas E, Fish J, Chai B, Farris RJ, Kulam-Syed-Mohideen AS, McGarrell DM, Marsh T, Garrity GM, and Tiedje JM. 2009. The Ribosomal Database Project: improved alignments and new tools for rRNA analysis. *Nucleic Acids Res.* 37 (Database issue): D141-D145.

Dean FB, Nelson JR, Giesler TL, and Lasken RS. 2001. Rapid amplification of plasmid and phage DNA using Phi 29 DNA polymerase and multiply-primed rolling circle amplification. *Genome Res.* 11 (6): 1095-1099.

Desantis TZ, Hugenholtz P, Larsen N, Rojas M, Brodie EL, Keller K, Huber T, Dalevi D, Hu P, and Andersen GL. 2006. Greengenes, a chimera-checked 16S rRNA gene database and workbench compatible with ARB. *Appl. Environ. Microbiol.* 72 (7): 5069-5072.

Dixon TC, Fadl AA, Koehler TM, Swanson JA, and Hanna PC. 2000. Early *Bacillus anthracis*-macrophage interactions: intracellular survival survival and escape. *Cell Microbiol.* 2 (6): 453-463.

Dowd SE, Wolcott RD, Sun Y, McKeehan T, Smith E, and Rhoads D. 2008. Polymicrobial nature of chronic diabetic foot ulcer biofilm infections determined using bacterial tag encoded FLX amplicon pyrosequencing (bTEFAP). *PLoS One.* 3 (10): e3326.

Dressman D, Yan H, Traverso G, Kinzler KW, and Vogelstein B. 2003. Transforming single DNA molecules into fluorescent magnetic particles for detection and enumeration of genetic variations. *Proc Natl Acad Sci U. S. A* 100 (15): 8817-8822.

Fedurco M, Romieu A, Williams S, Lawrence I, and Turcatti G. 2006. BTA, a novel reagent for DNA attachment on glass and efficient generation of solid-phase amplified DNA colonies. *Nucleic Acids Res.* 34 (3): e22.

Gevers D, Cohan FM, Lawrence JG, Spratt BG, Coenye T, Feil EJ, Stackebrandt E, Van de PY, Vandamme P, Thompson FL, and Swings J. 2005. Opinion: Re-evaluating prokaryotic species. *Nat Rev. Microbiol.* 3 (9): 733-739.

Ghadessy FJ, Ong JL, and Holliger P. 2001. Directed evolution of polymerase function by compartmentalized self-replication. *Proc Natl Acad Sci U. S. A* 98 (8): 4552-4557.

Hamady M, and Knight R. 2009. Microbial community profiling for human microbiome projects: Tools, techniques, and challenges. *Genome Res.* 19 (7): 1141-1152.

Hamady M, Walker JJ, Harris JK, Gold NJ, and Knight R. 2008. Error-correcting barcoded primers for pyrosequencing hundreds of samples in multiplex. *Nat Methods* 5 (3): 235-237.

Handelsman J. 2004. Metagenomics: application of genomics to uncultured microorganisms. *Microbiol. Mol Biol. Rev.* 68 (4): 669-685.

Hongoh Y. 2010. Diversity and genomes of uncultured microbial symbionts in the termite gut. *Biosci. Biotechnol. Biochem.* 74 (6): 1145-1151.

Huson DH, Auch AF, Qi J, and Schuster SC. 2007. MEGAN analysis of metagenomic data. *Genome Res.* 17 (3): 377-386.

Klyachko NL, and Levashov AV. 2003. Bioorganic synthesis in reverse micelles and related systems. *Current Opinion in Colloid and Interface Science* 8 (2): 179-186.

Krause L, Diaz NN, Goesmann A, Kelley S, Nattkemper TW, Rohwer F, Edwards RA, and Stoye J. 2008. Phylogenetic classification of short environmental DNA fragments. *Nucleic Acids Res.* 36 (7): 2230-2239.

Lane DJ, Pace B, Olsen GJ, Stahl DA, Sogin ML, and Pace NR. 1985. Rapid determination of 16S ribosomal RNA sequences for phylogenetic analyses. *Proc Natl Acad Sci U. S. A* 82 (20): 6955-6959.

Liu P, and Mathies RA. 2009. Integrated microfluidic systems for high-performance genetic analysis. *Trends in Biotechnology* 27 (10): 572-581.

Liu Z, Lozupone C, Hamady M, Bushman FD, and Knight R. 2007. Short pyrosequencing reads suffice for accurate microbial community analysis. *Nucleic Acids Res.* 35 (18): e120.

Liu Z, DeSantis TZ, Andersen GL, and Knight R. 2008. Accurate taxonomy assignments from 16S rRNA sequences produced by highly parallel pyrosequencers. *Nucleic Acids Research* 36 (18): e120.

Maiden MC. 2006. Multilocus sequence typing of bacteria. *Annu. Rev. Microbiol.* 60: 561-588.

Mardis ER. 2011. A decade's perspective on DNA sequencing technology. *Nature* 470 (7333): 198-203.

Margulies M, Egholm M, Altman WE, Attiya S, Bader JS, Bemben LA, Berka J, Braverman MS, Chen YJ, Chen Z, Dewell SB, Du L, Fierro JM, Gomes XV, Godwin BC, He W, Helgesen S, Ho CH, Irzyk GP, Jando SC, Alenquer ML, Jarvie TP, Jirage KB, Kim JB, Knight JR, Lanza JR, Leamon JH, Lefkowitz SM, Lei M, Li J, Lohman KL, Lu H, Makhijani VB, McDade KE, McKenna MP, Myers EW, Nickerson E, Nobile JR, Plant R, Puc BP, Ronan MT, Roth GT, Sarkis GJ, Simons JF, Simpson JW, Srinivasan M, Tartaro KR, Tomasz A, Vogt KA, Volkmer GA, Wang SH, Wang Y, Weiner MP, Yu P, Begley RF, and Rothberg JM. 2005. Genome sequencing in microfabricated high-density picolitre reactors. *Nature* 437 (7057): 376-380.

Mellmann A, Harmsen D, Cummings CA, Zentz EB, Leopold SR, Rico A, Prior K, Szczepanowski R, Ji Y, Zhang W, McLaughlin SF, Henkhaus JK, Leopold B, Bielaszewska M, Prager R, Brzoska PM, Moore RL, Guenther S, Rothberg JM, and Karch H. 2011. Prospective genomic characterization of the German enterohemorrhagic *Escherichia coli* O104:H4 outbreak by rapid next generation sequencing technology. *PLoS One.* 6 (7): e22751.

Metzker ML. 2010. Sequencing technologies - the next generation. *Nat. Rev. Genet.* 11 (1): 31-46.

Meyer F, Paarmann D, D'Souza M, Olson R, Glass EM, Kubal M, Paczian T, Rodriguez A, Stevens R, Wilke A, Wilkening J, and Edwards RA. 2008. The metagenomics RAST server - a public resource for the automatic phylogenetic and functional analysis of metagenomes. *BMC. Bioinformatics.* 9: 386.

Mullis K, Faloona F, Scharf S, Saiki R, Horn G, and Erlich H. 1986. Specific enzymatic amplification of DNA *in vitro*: the polymerase chain reaction. *Cold Spring Harb. Symp. Quant. Biol.* 51 Pt 1: 263-273.

Mullis KB, and Faloona FA. 1987. Specific synthesis of DNA *in vitro* via a polymerase-catalyzed chain reaction. *Methods Enzymol.* 155: 335-350.

Neefs JM, Van de PY, De RP, Chapelle S, and De WR. 1993. Compilation of small ribosomal subunit RNA structures. *Nucleic Acids Res.* 21 (13): 3025-3049.

Ottesen EA, Hong JW, Quake SR, and Leadbetter JR. 2006. Microfluidic digital PCR enables multigene analysis of individual environmental bacteria. *Science* 314 (5804): 1464-1467.

Pace NR. 1997. A molecular view of microbial diversity and the biosphere. *Science* 276 (5313): 734-740.

Raghunathan A, Ferguson HR, Jr., Bornarth CJ, Song W, Driscoll M, and Lasken RS. 2005. Genomic DNA amplification from a single bacterium. *Appl. Environ. Microbiol.* 71 (6): 3342-3347.

Sanger F, Nicklen S, and Coulson AR. 1977. DNA sequencing with chain-terminating inhibitors. *Proc Natl Acad Sci U. S. A* 74 (12): 5463-5467.

Sapkota AR, Berger S, and Vogel TM. 2010. Human pathogens abundant in the bacterial metagenome of cigarettes. *Environ. Health Perspect.* 118 (3): 351-356.

Shendure J, Porreca GJ, Reppas NB, Lin X, McCutcheon JP, Rosenbaum AM, Wang MD, Zhang K, Mitra RD, and Church GM. 2005. Accurate multiplex polony sequencing of an evolved bacterial genome. *Science* 309 (5741): 1728-1732.

Shendure JA, Porreca GJ, Church GM, Gardner AF, Hendrickson CL, Kieleczawa J, and Slatko BE. 2011. Overview of DNA Sequencing Strategies. In *Current Protocols in Molecular Biology*, John Wiley & Sons, Inc.

Shwed P.S., Crosthwait J, and Seligy VL. 2007. Comparative genomic and DNA microarray applications in risk assessment of biotechnology microorganisms. In *Recent Advancements in Analytical Biochemistry: Applications in Environmental Toxicology*, eds P Kumarathasan and R Vincent, 137-156. Research Signpost.

Stahl DA, Lane DJ, Olsen GJ, and Pace NR. 1984. Analysis of hydrothermal vent-associated symbionts by ribosomal RNA sequences. *Science* 224 (4647): 409-411.

Stahl DA, Lane DJ, Olsen GJ, and Pace NR. 1985. Characterization of a Yellowstone hot spring microbial community by 5S rRNA sequences. *Appl. Environ. Microbiol.* 49 (6): 1379-1384.

Sykes PJ, Neoh SH, Brisco MJ, Hughes E, Condon J, and Morley AA. 1992. Quantitation of targets for PCR by use of limiting dilution. *Biotechniques* 13 (3): 444-449.

Szita N, Polizzi K, Jaccard N, and Baganz F. 2010. Microfluidic approaches for systems and synthetic biology. *Curr. Opin. Biotechnol.* 21 (4): 517-523.

Tadmor AD, Ottesen EA, Leadbetter JR, and Phillips R. 2011. Probing individual environmental bacteria for viruses by using microfluidic digital PCR. *Science* 333 (6038): 58-62.

Tewhey R, Warner JB, Nakano M, Libby B, Medkova M, David PH, Kotsopoulos SK, Samuels ML, Hutchison JB, Larson JW, Topol EJ, Weiner MP, Harismendy O, Olson J, Link DR, and Frazer KA. 2009. Microdroplet-based PCR enrichment for large-scale targeted sequencing. *Nat Biotechnol.* 27 (11): 1025-1031.

Vandamme P, Pot B, Gillis M, De VP, Kersters K, and Swings J. 1996. Polyphasic taxonomy, a consensus approach to bacterial systematics. *Microbiol. Rev.* 60 (2): 407-438.

von Mering C, Hugenholtz P, Raes J, Tringe SG, Doerks T, Jensen LJ, Ward N, and Bork P. 2007. Quantitative phylogenetic assessment of microbial communities in diverse environments. *Science* 315 (5815): 1126-1130.

Woese CR. 1987. Bacterial evolution. *Microbiol. Rev.* 51 (2): 221-271.

Woyke T, Tighe D, Mavromatis K, Clum A, Copeland A, Schackwitz W, Lapidus A, Wu D, McCutcheon JP, McDonald BR, Moran NA, Bristow J, and Cheng JF. 2010. One bacterial cell, one complete genome. *PLoS One.* 5 (4): e10314.

Yang B, Peng Y, Leung HC, Yiu SM, Chen JC, and Chin FY. 2010. Unsupervised binning of environmental genomic fragments based on an error robust selection of l-mers. *BMC. Bioinformatics.* 11 Suppl 2: S5.

Zhang K, Martiny AC, Reppas NB, Barry KW, Malek J, Chisholm SW, and Church GM. 2006. Sequencing genomes from single cells by polymerase cloning. *Nat Biotechnol.* 24 (6): 680-686.

# 9

# Submicroscopic Human Parasitic Infections

Fousseyni S. Touré Ndouo
*Medical Parasitology Unit,*
*Centre International de Recherches Médicales de Franceville (CIRMF), Franceville*
*Gabon*

## 1. Introduction

Polymerase chain reaction (PCR) amplification provides a powerful tool for parasite detection. This chapter examines the use of PCR to diagnose malaria in patients with low parasite densities (submicroscopic infections, SMI) and also occult loaiosis (OL: *Loa loa* infection without detectable circulating microfilaria on standard microscopy). It provides therefore the issue of management of these kinds of infections with regard to the eradication policy of such pathogens.

### 1.1 Classification

i.  **Malaria:** Malaria is caused by *Plasmodium* parasites, of which there are about 200 species (Levine ND 1980). These protozoans belong to the *Apicomplexa* phylum, *Sporozoa* class and *Haemosporidae* subclass (Levine ND 1970). They are obligatory intracellular parasites. Two successive hosts, humans and mosquitoes (Culicidea and Anphelinea), are necessary for their life cycle. Four main species infect humans, namely *Plasmodium falciparum*, *P. vivax*, *P. malariae* and *P. ovale*. A fifth species, *P. knowlesi*, is currently spreading in south-east Asia and Oceania. This species derived from chimpanzees has caused more than 250 human cases of malaria in Malaysia but is still considered to be zoonotic (Figtree et al. 2010). *P. falciparum* causes most life-threatening infections.
Human is the intermediate host for malaria, wherein the asexual phase of the life cycle occurs. The sporozoïtes, inoculated by the infested female *Anopheles* mosquito, initiate this phase of the cycle from the liver, and continue within the red blood cells. From the mosquito bite, tens to few hundred invasive sporozoïtes are introduced into the skin. Following the intradermal drop, some sporozoïtes are destroyed locally by the immune cells, or enter into the lymphatic vessels, and some others can find blood circulation (Megumi L et al. 2007; Ashley M et al. 2008; Olivier S et al. 2008). The sporozoïtes that find peripheral blood circulation reach the liver within a few hours. It has been recently demonstrated that these sporozoïtes travel by a continuous sequence of stick-and-slip motility, using the thrombospondin-related anonymous protein (TRAP) family and an actin–myosin motor (Baum J et al. 2006; Megumi L et al. 2007; Münter S et al. 2009). The sporozoites migrate into hepatocytes and then grow within parasitophorous vacuoles and develop to the schizont stage which releases merozoites (Jones MK et al. 2006; Kebaier C et al. 2009). The entire pre-eryhrocytic phase lasts about 5–16 days depending

on the parasite species (5-6 days for *P. falciparum*, 8 days for *P. vivax*, 9 days for *P. ovale*, 13 days for *P. malariae* and 8-9 days for *P. knowlesi*. The pre-erythrocytic phase remains a "silent" phase, with little pathology and no symptoms, as only a few hepatocytes are affected (Ashley M et al. 2008). This phase is a single cycle, contrasting to the next, erythrocytic stage, which occurs repeatedly.

ii. **Loaiosis:** Filariasis are typically chronic tropical diseases caused by nematodes of the *Filariidae* family, transmitted by flies or mosquitoes. Eight species are currently known to infect humans, namely *Wuchereria bancrofti*, *Brugia malayi*, *B. timoni*, *Onchocerca volvulus*, *Loa loa*, *Mansonella perstans*, *M. ozzardi*, and *M. streptocerca*. Three groups of filariasis have been distinguished on the basis of their human target tissues: lymphatic filariasis (wuchereriasis and brugiasis); cutaneous dermal filariasis (loaiosis, onchocerciasis, and *streptocerca* mansonelliasis) and serous filariasis (*perstans* and *ozzardi* mansonelliasis) (Gentilini 1982). The vectors are blood-sucking flies and female mosquitoes. The microfilarial eggs or embryos are ingested by the vector when it bites an infected human. These microfilariae become infective stage L3 larvae after two successive mutes within the vector, and are transmitted to a new human host through a new blood meal or bite.

More than 3.3 billion people are exposed to filariosis, and an estimated 300 million people are infected. Loaiosis occurs in Africa, brugiaioses in South Asia, wuchereriasis in Africa and Asia, onchocerciasis in Africa, Central and South America and Asia (Yemen), *perstans* mansonelliasis in Africa and Central and South America, *ozzardi* mansonelliasis in Central and South America, and *streptocerca* mansonellaiosis in Africa.

*L. loa* infection (loaiosis) was initially described in 1770 by Mongin, in a female slave originating from West Africa and living on Saint Domingue island. Guyon et al. found the same worm in Gabon (Central Africa) in 1864. *L. loa* was first described in detail by Brumpt *et al.* in 1904, and then by Connors *et al.* in 1976. Although discovered in the Antilles, *L. loa* is restricted to Africa (Gentilini 1982). The adult worms live under the skin for about 15 years (Gentilini et al. 1982). The tabanides responsible for *L. loa* transmission are primarily *Chrysops dimidiata* and *silacea*, two forest species often present in the same hearth. Only the females are hematophagous, and they have diurnal activity.

## 1.2 Pathogenesis

i. **Malaria:** *P. falciparum* is responsible for most complicated forms of malaria and causes about 800 000 deaths a year, mostly among children in sub-Saharan African countries (WHO 2009). Malaria symptoms generally occur in three phases. After an incubation period of 7 to 10 days, symptoms begin with fever, aches and digestive disorders (febrile stomach upset). Then, when schizont rupture becomes synchronous, patients enter the feverish reviviscent schizogonic phase (periodic fever) of uncomplicated malaria. This phase is characterized by fevers typically appearing every 24 hours (third fever in infection by *P. vivax* or *ovalae*, every 48 hours, quartan fever in infection by *P. malariae* or *P. falciparum*), accompanied by a triad of symptoms: shivers, fever and sweating. Destruction of parasitized red blood cells leads to the release of malarial toxins and to TNF alpha production. The third phase, mainly seen with *P. falciparum*, corresponds to severe malaria (pernicious access), which sometimes occurs rapidly after infection. Clinical and biological signs are used to classify malaria (WHO 2000 gravity criteria, Imbert et al 2002). The reasons why some non immune individuals infected by *P. falciparum* develop severe malaria and die, while others have only uncomplicated malaria or remain asymptomatic, remain unclear (Marsh et al 1988). Severe anemia and cerebral malaria are responsible for most of the morbidity and mortality related to this disease in children. Despite abundant research, the pathophysiological mechanisms underlying severe forms are poorly understood. Several studies have implicated sequestration of *P. falciparum*-parasitized red blood cells (PRBC) in the lungs and brain (Taylor et al. 2001). This sequestration is characterized by PRBC adhesion (or cytoadherence), agglutination and rosetting. Cytoadherence of PRBC to host endothelial cells (EC) in brain and lung capillaries can obstruct the microvasculature, a phenomenon accompanied by changes in the T cell repertoire and by cytokine production (Mazier et al. 2000). This adherence is modulated by platelets (Brown et al. 2000) and is mediated by EC receptors such as CD36, intracellular adhesion molecule 1 (ICAM1), vascular cellular adhesion molecule 1 (VCAM1), CD31, integrins and hyaluronic acid (Hunt et al. 2003). PRBC adhesion can induce over-expression of inflammatory cytokines (Mazier et al. 2000) and EC apoptosis (Pino et al. 2003). Approximately 20% of *P. falciparum* isolates from Franceville, Gabon (Central Africa), were show to induce human lung endothelial cell (HLEC) apoptosis by cytoadherence (Touré et al. 2008). In addition, apoptogenic isolates were more frequent in children with neurological signs (prostration or coma), supporting the hypothesis that PRBC-mediated EC apoptosis could amplify blood-brain barrier disruption and dysfunction (Combes et al. 2005; Bisser et al. 2006). Whole transcriptome analysis revealed that 59 genes were more intensely transcribed in apoptogenic strains than in non apoptogenic strains (Siau et al. 2007). Knock-down of 8 of these genes by double-strand RNA interference significantly reduced the apoptogenic response in 5 genes (PF07_0032, PF10255, PFI0130c, PFD0875c, and MAL13P1.206). These five genes are known as *Plasmodium* apoptosis-linked pathogenicity factors (PALPF).

ii. **Loaiosis:** Loaiosis is characterized by calabar oedema (swelling) and conjunctivitis due to ocular passage of adult worms. Calabar oedema is transient and located on the face, limbs and back of the hands and fingers. *L. loa* is also called the "African eye worm". Meningoencephalic complications are an adverse effect of diethylcarbamazine treatment for hypermicrofilaremia. Other complications such as nephropathies, endocarditis, retinopathies, neuropathies and pneumonitis have been reported (Schofield et al. 1955; Hulin et al. 1994). Symptoms are more frequent in expatriates. Immunologically, loaiosis is characterized by hypergammaglobulinemia, hypereosinophilia and high IgE levels responsible for allergic symptoms. In endemic areas, loaiosis is the third reason for medical consultations in rural settings, although many microfilaremic subjects are asymptomatic. Occult loaiosis (amicrofilaremic infection) is defined as infection by the adult worm without peripheral microfilaremia on standard microscopy. Amicrofilaremic status is common among autochthonous residents and may be due to sequestration of microfilaria or to their massive destruction by the immune system, or to the presence of sterile adult worms. This form of infection is the most common in endemic areas. Other amicrofilaremic subjects are thought to be resistant. There is currently no way of discriminating between these two amicrofilaremic subgroups in the absence of (transient) ocular passage of adult worms.

## 1.3 Diagnostic challenges

i.   **Malaria:** Light microscopy of blood smears remains the standard method for *Plasmodium* detection, both for clinical diagnosis and epidemiological surveys (Okell LC et al. 2009). However, sensitivity depends on parasite density in blood. In patients with low parasitemia, mixed infections, antimalarial treatment or chronic infection, microscopic diagnosis requires painstaking examination by an experienced technician. Low-density infections that cannot be detected by conventional microscopy are termed submicroscopic infections (SMI). *Plasmodium* species identification is mainly based on microscopic morphological characteristics but this is not entirely reliable (*Plasmodium vivax* resembles *P. ovale*). In addition, parasite morphology can be altered by drug treatment and/or sample storage conditions.

ii.  **Loaiosis:** Human loaiosis differs from other filariasis by the fact that most patients have "occult" infection, with no circulating microfilaria. This peripheral amicrofilaremia can be due to microfilaria destruction by the immune system, and/or to their sequestration. These subjects cannot be diagnosed by microscopy and consequently go untreated, constituting a parasite reservoir. Before 1997, *L. loa* diagnosis was still based on microscopic examination and the prevalence was therefore underestimated. In contrast, because of cross-reactions, serological tests, and especially those based on total IgG detection, tend to overestimate prevalence. The existence of many cases of occult but symptomatic infection among residents in endemic areas implies the need for specific and sensitive detection.

# 2. PCR-based diagnosis of malaria

In 1993 a PCR method targeting the small subunit of the ribosomal RNA (SSUrRNA) gene was developed for use as an alternative to microscopy for detecting the four main *Plasmodium* species (Snounou et al. 1993, 1994, 1995). Nested PCR was used for its high sensitivity and specificity (Snounou et al. 1993). However, the nested reaction requires five separate PCR reactions and is therefore time-consuming, expensive and not always feasible in developing world laboratories. Several variants of this nested PCR method, such as semi-nested multiplex and one-tube multiplex have been developed (Mixon-Hayden T et al. 2010). In 1998 Jarra and Snounou showed that *Plasmodium* DNA is cleared very quickly from the bloodstream and that positive PCR amplification is usually associated with the presence of viable parasites. PCR positivity therefore indicates active *Plasmodium* infection. Since 1997, several PCR methods targeting other *P. falciparum* genes have been developed (Cheng et al. 1997; Filisetti et al. 2002). Their sensitivity has been estimated at 71%, 83% and 100% for the *MSP-2*, *SSUrRNA* and *STEVOR* genes, respectively (Oyedeji et al. 2007).

Real-time PCR has been reported to be able to improve parasite detection. Compared to *SSUrRNA* nested PCR, the real-time assay had a sensitivity of 99.5% and specificity of 100% for the diagnosis of malaria (Farcas GA et al. 2004). The real-time PCR method, specific for all *Plasmodium* species, avoids post-amplification sample handling and electrophoresis, and the result can be ready within 45 min (Farcas GA et al. 2004). This method would be useful for monitoring antimalarial drug efficacy, especially in areas of drug resistance (Lee MA et al. 2002).

More recently, it has been shown that dot18S (18SrRNA gene) and CYTB, two new molecular methods, are highly sensitive and allow high-throughput scaling up for many hundred samples (Steenkeste N et al. 2009). The CYTB is a nested PCR based on Plasmodium cytochrome b gene followed by species detection using SNP (single nucleotide polymorphism) analysis. The usefulness of these methods in detecting malaria has been demonstrated especially in low endemic areas.

## 2.1 Materials and methods

a.  **Blood sampling**
    Samples must be collected in sterile tubes. For example, peripheral blood is collected in tubes containing an anticoagulant such as EDTA. However, some anticoagulants, such as heparin, inhibit the action of Taq DNA polymerase and should thus be avoided. Blood samples can also be collected in the form of drops on calibrated pre-punched paper disks (Serobuvard, LDA 22H, Zoopole, Ploufragan, France) (Ouwe-Missi-Oukem-Boyer et al. 2005).

b.  **Microscopy**
    Thick and thin peripheral blood films were stained with Giemsa and examined by microscope. For microscope positive samples, the parasite load is expressed as the number of asexual forms of *P. falciparum*/µL of blood, assuming an average leukocyte count of 8000/µL.

c.  **DNA template preparation**
    There are many useful techniques for DNA template processing. Plasmodial DNA extraction involves erythrocyte lysis and proteinase K digestion to prevent PCR inhibition.

    i.   **DNeasy[R] Blood & Tissue Kit:** Whole blood (200 µl) is used for DNA extraction with the DNeasy Blood & Tissue kit (QIAGEN, Hilden, Germany). Briefly, DNA extraction is carried out as follows. To a 1.5-ml tube containing 200 µl of while blood are added 20 µl of proteinase K solution and 200 µl of AL buffer (a detergent included in the kit). The mixture is pulse vortexed for 15 seconds and incubated for 15 minutes at 56°C. Two hundred microliters of cold ethanol is then added and the mixture is vortexed for 15 seconds. The mixture is transferred to a mini-column assembled on a 2-ml tube and centrifuged for 1 min at 8000 rpm. After centrifugation the 2-ml tube is discarded OK. The mini-column is recovered and placed on a new 2-ml tube. The mini-column is then washed with 500 µL of AW1 buffer (available in the kit) by centrifugation at 8000 rpm for 1 min. This washing step is repeated with another 500 µL of AW2 buffer, followed by centrifugation for 3 min at 14 000 rpm. The 2-ml tube is again discarded.
         The mini-column is placed on a 1.5-ml tube and 60 µl of AE elution buffer is added. This unit is left at room temperature for 10 min and then centrifuged for 1 min at 8000 rpm. The DNA is then recovered in the 1.5-ml tube and immediately used as a template or stored at 20°C.

    ii.  **Dried blood-spot method** (DBS): DNA templates are extracted as described by Plowe CV et al in 1995. The dried blood spot is placed in 1 ml of phosphate buffered saline (PBS) containing 0.5% saponin and is incubated overnight at 4°C. The resulting brown solution is replaced with 1 ml of PBS and incubated for an additional 15-30 minutes at 4°C. Then, 200 µl of 5% Chelex 100 (Bio-Rad Laboratories, CA) is placed in clean tubes and heated to 100°C in a water bath. The disks are removed from the PBS and placed in the preheated 5% Chelex 100, vortexed at high speed for 30 seconds and placed in a water bath at 100°C for 10 minutes with gentle agitation. The samples are then centrifuged at 10 000 *g* for 2 minutes, and the supernatant is removed and centrifuged as before. The supernatant is then collected in a clean tube and immediately used for PCR or stored at 20°C until use.

DNA can be also extracted from dried blood spots with several other methods, such as the QIAamp® DNA Mini Kit (QIAGEN, Hilden, Germany).

### 2.2 *P. falciparum* DNA amplification and detection

#### i.   SSUrRNA gene amplification

Two microliters of DNA extract is amplified in a final volume of 25 µl containing 2.5 µl of 10X reaction buffer, 100 µM each dNTP (dATP, dGTP, dTTP, and dCTP), 0.5 pM each primer (rPLU5/rPLU6 (rPLU5 5'-CCT GTT GTT GCC TTA AAC TTC-3' and rPLU6 5'-TTA AAA TTG TTG CAG TTA AAA CG-3') for the primary reaction, and rFAL1/rFAL2 (rFAL 1 5'-TTA AAC TGG TTT GGG AAA ACC AAA TAT ATT-3' and rFAL 2 5'-ACA CAA TGA ACT CAA TCA TGA CTA CCC GTC-3') for the nested reaction) and 0.75 units of Taq DNA polymerase (QIAGEN, Hilden, Germany). The primer sequences (Table 1) are based on SSUrRNA sequences described elsewhere (Snounou et al. 1993). The PCR program is as follows: denaturation at 95°C for 5 min followed by 25 cycles (30 cycles in nested PCR) at 94°C for 1 min, 60°C for 2 min and 72°C for 2 min, with a final extension step of 5 min at 72°C.

#### ii.   STEVOR gene amplification

The first round of amplification is performed with a reaction mix of 50 µl containing 5.0 µl of 10X reaction buffer, 200 µM each dNTP (dATP, dGTP, dTTP, and dCTP), 1.25 units of Taq

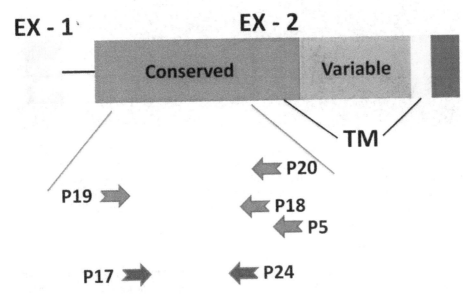

Schema 1. Schematic representation of the STEVOR PCR methodology (CHENG et al 1997).

DNA polymerase, 0.4 pM each primer (P5, P18, P19 and P20) (P5 5'-GGG AAT TCT TTA TTT GAT GAA GAT G-3', P18 5'-TTT CA(C/T) CAC CAA ACA TTT CTT-3', P19 5'-AAT CCA CAT TAT CAC AAT GA-3', P20 5'-CCG ATT TTA ACA TAA TAT GA-3') and 5 μl of DNA template. The PCR program is as follows: denaturation at 93°C for 3 min followed by 25 cycles of 30 s at 93°C, 50 s at 50°C and 30 s at 72°C, with a final extension step of 3 min at 72°C. Two microliters of the first-round PCR product is used for the second round of amplification, with a reaction mixture of 50 μl containing 5.0 μl of 10X reaction buffer, 200 μM each dNTP, 1.25 units of Taq DNA polymerase and 0.4 pM each primer (P17 and P24) (P17 5'-ACA TTA TCA TAA TGA (C/T) CC AGA ACT-3', P24 5'-GTT TGC AAT AAT TCT TTT TCT AGC-3'). The PCR conditions for the nested reaction are as follows: denaturation at 93°C for 3 s, followed by 25 cycles of 30 s at 93°C, 50 s at 55°C and 30 s at 72°C, with a final extension step of 3 min at 72°C.

### iii. Detection procedures

**Analysis of PCR products:** After amplification, 10 μl of each PCR product is mixed with 1 μl of loading dye (0.25% bromophenol blue, 0.25% xylene cyanol and 40% w/v sucrose in water) and analyzed by electrophoresis on 1.5% agarose gel. The gel is stained with ethidium bromide or FluoProbes Gel Red (Interchim Montlucon, France) and the DNA is visualized and photographed under ultraviolet light.

*Plasmodium* **SSUrRNA gene:**

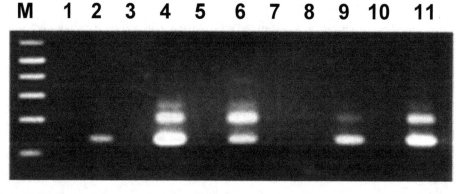

Fig. 1. Detection and speciation of *Plasmodium* by nested PCR using genus-specific primers and 1.5% agarose gel electrophoresis. Lanes 1 and 8: PCR-negative controls; lane 2: an individual with submicroscopic infection by *P. malariae* (size: 144 base pairs); lanes 3, 5, 7 and 10: PCR-negative individuals; lanes 4, 6, 9 and 11: individuals with submicroscopic co-infection with *P. falciparum* (size: 205 bp) and *P. malariae*. Lane M represents the DNA molecular weight marker (100 bp).

### *P. falciparum* **STEVOR gene**

Theoretically, three specific bands between 189-700 base pairs are generated using nested primers. We obtained a specific band of 250 bp for all *P. falciparum* isolates tested in Franceville, southeastern Gabon.

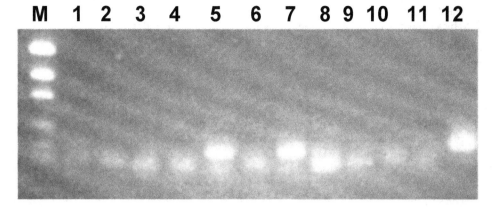

Fig. 2. Detection of the *Plasmodium* STEVOR gene by nested PCR using specific primers and 1.5% agarose gel electrophoresis. Lanes 1 and 10: PCR-negative controls; lanes 2, 3, 4, 6, 8, 9 and 11: PCR-negative samples; lanes 5 and 7: PCR-positive samples; lane 12: PCR-positive control; lane M: DNA molecular weight marker (123 bp).

## 3. PCR based diagnosis of *Loa loa*

In 1997, a PCR method (15r3-PCR) was developed to detect the repeat 3 region of the gene encoding the *L. loa* polyprotein in blood samples (Touré et al. 1997a, 1997b). Amicrofilaremic status is generally due to massive destruction of microfilaria, releasing parasite DNA into the bloodstream. These molecules may exist free in plasma, or be associated with cell-surface proteins, or even be contained in phagocytic cells. In addition, the adult worms can release DNA when they produce nonviable eggs or when they die after immune attack. The quantity of DNA released, whether from eggs, microfilaria or adult worms, is related to the parasite load of adult worms. The 15r3-PCR assay had a sensitivity of 95% with respect to detection of ocular passage of *L. loa* adult worms, and 100% compared to detection of microfilaremia.

### 3.1 Materials and methods

a.  Blood sampling
    As previously mentioned blood samples must be collected by venipuncture into Vacutainer tubes containing an adequate anticoagulant such as EDTA.

b.  Leukoconcentration
    Clinically, *L. loa* infection is diagnosed when migration of adult worms under the conjunctiva and/or skin is observed, or when a patient presents with classical symptoms. Diagnosis is classically based on standard microscopy. Microfilariae are the blood stage of *L. loa*. One milliliter of each blood sample is added to a 15-ml tube containing 9 ml of phosphate buffered saline (PBS), in duplicate. The mixture is treated with 600 μl of 2% saponin at room temperature for 15 min to lyse red blood cells, followed by centrifugation at 1000 *g* for 15 minutes at 4°C. The supernatant is discarded and the pellet is examined microscopically for microfilariae. The distinction between *L.*

*loa* and *M. perstans* microfilariae is based on size, motility, and by the presence of a sheath (*L. loa*).

Thick smears can be also prepared with venous blood and stained with Giemsa or hematoxylin-eosin to detect microfilariae. The QBC (Quantitative Coating Buffer) method is also used to detect *L. loa* microfilariae.

For microscopic detection of *L. loa* microfilariae, blood samples must be collected during the day, given the diurnal periodicity of human loaiosis.

c.   Whole blood lysate processing
     Whole blood (100 µl) is mixed with 500 µl of TE buffer (10 mM Tris pH 8; 0.1 mM EDTA pH 8) and spun at 10 000 *g* twice for 2 min, discarding the supernatant at each step. The pellet is resuspended in 500 µl of sucrose buffer (10 mM Tris pH 7.6, 5 mM MgCl₂ 1 M sucrose and 1% Triton X 100) and spun at 10 000 *g* twice for 2 min. After the final wash, the supernatant is discarded and the pellet is resuspended with 200 µl of prewarmed (56°C) proteinase K buffer (containing 20 mM Tris pH 8, 50 mM KCl, 2.5 mM MgCl₂, 100 µg/ml proteinase K and 0.5% Tween20), incubated at 56°C for two hours, then held at 90°C for 10 min. The DNA can be stored at 4°C for several days or at -20°C until required.

d.   *L. loa* 15r3 gene amplification and detection
     Primers corresponding to the 5' and 3' ends of the repeat 3 sequence of the gene coding for *L. loa* 15 kDa allergenic polyprotein are used. Primary amplification is done with a reaction mixture of 50 µl containing 2 µl of blood lysate, 1X PCR buffer (supplied by the manufacturer: 200 mM Tris-HCl pH 8.7, 100 mM KCl, 100 mM (NH₄)₂SO₄, 20 mM MgSO₄, 1% Triton x100, 1 mg/ml bovine serum albumin), 200 µM each dATP, dCTP, dGTP and dTTP, 1 µM each primer (15r3-1: 5'-AAT-CAG-GCA-AAT-AAT-GGC-ACA-AAA-3', 15r3-2: 5'-GCG-TTT-TCT-TCT-CAC-CAG-CTG-TCT-3') and 1 unit of DNA polymerase. Amplification is performed with a Perkin Elmer thermal cycler for 40 cycles: 94°C for 1 min (denaturation), 65°C for 1 min (annealing) and 72°C for 2 min

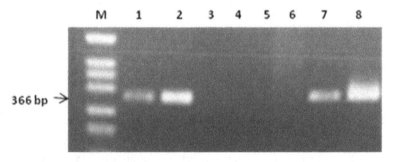

Fig. 3. Representative 1.5% agarose gel electrophoresis patterns of nested 15r3 PCR products. Lanes 1, 2 and 7: *L. loa* amicrofilaremic individuals (AMF) positive by PCR. Lanes 4 and 5: individuals negative by PCR. Lane 8: an individual with 100 *L. loa* microfilariae per ml and positive by PCR. Lanes 3 and 6: PCR-negative controls (no template); 5 µl of each nested PCR product was applied to each lane and revealed using UV transillumination after ethidium bromide staining. A fragment of 366 bp was observed with positive samples. Lane M: DNA molecular weight marker VI (Boehringer).

(extension), preceded by a "hot start" cycle at 96°C for 10 min, 80°C for 5 min and 94°C for 30 s. One microliter of product from the first-round amplification is used for a second round in the above conditions for 30 cycles. The following primers are used: 15r3-3: 5'GGC ACA AAA CAC TGC AGC AGT CCT3', and 15r3-4: 5'CAG CTG TCT CAA ATC GAA GAT TCT 3.'

## 4. Submicroscopic infection and disease management and control

### i. Malaria:

The global strategy for malaria control is based on prevention, early diagnosis and prompt treatment. The detection limit of routine microscopy has been estimated to be about 100 parasites/milliliter, whereas PCR can detect as little as 0.01 parasite /micro liter (Mockenhaupt FP et al. 2002). Submicroscopic infection (SMI) including submicroscopic gametocytes is common in both symptomatic and asymptomatic individuals with malaria. A systematic review and analysis of field data carried out by Okell LC et al. in 2009 showed that the prevalence of *P. falciparum* was twice as high with PCR as with microscopy. In a village in Dienga, southeastern Gabon, PCR was performed on blood samples from asymptomatic individuals negative by microscopy: the prevalence of SMI (PCR positivity) was 13.7% by PCR and 7.2% by microscopy (Touré et al. 2006). A study carried out by Bouyou-Akotet et al. in 2010 in Libreville (capital of Gabon) showed an 18.2% prevalence of SMI in pregnant women. Recently, SMI was detected in 18% of symptomatic individuals in Franceville, southeastern Gabon, whereas the microscopic prevalence was 23% (author's personal data). It has been estimated that as many as 88% of infections remain undetectable by microscopy in low-transmission areas, where the PCR prevalence is generally under 10% (Okell LC et al. 2009). Thus, a high rate of SMI could undermine disease control programs. In endemic areas, it has been shown that *P. falciparum* SMI contributes to acute disease (Rogier C et al. 1996), and to malaria-associated anemia and inflammation (Mockenhaupt FP et al. 2002). It has also been shown that cerebral malaria is frequently associated with SMI in semi-immune individuals (Giha HA et al. 2005). Finally, Bouyou-Akotet et al. 2010 have demonstrated that SMI during pregnancy is associated with low birth weight, especially in primagravidae. As parasite resistance to antimalarial drugs is currently widespread and increasing, it is very important to identify resistant parasites in patients with SMI. Two major genes have been implicated in *P. falciparum* resistance to quinoline, namely *Pfcrt* (*P. falciparum* chloroquine resistance transporter) and *Pfmdr1* (*P. falciparum* multidrug resistance gene 1). Single-nucleotide polymorphisms (SNPs) in these genes are associated with resistance both *in vitro* and *in vivo* (Wongsrichanalai et al. 2002). Therefore, *P. falciparum* drug resistance is linked to particular parasite genotypes (Duraisingh et al. 1997). *P. falciparum* infection is generally polyclonal, and may thus involve both drug-sensitive and resistant genotypes. SMI detection can be used to evaluate the therapeutic effectiveness of anti-malarial drugs during mass treatments and preclinical trials.

SMI individuals are capable of infecting mosquitoes and contributing to human transmission (Coleman RE et al. 2004), mainly in areas of seasonal transmission (Nwakanma D et al. 2008). Microscopy fails to detect the parasite in 49.2% of all malaria cases and in 91.3% of gametocytemic individuals (Okell LC et al. 2009). Individuals whose blood smears are negative for gametocytes (submicroscopic gametocyte) are equally able to transmit the infection to mosquitoes as slide-positive individuals (Coleman RE et al. 2004). Thus, the SMI gametocyte reservoir may sustain malaria transmission despite efforts to fight malaria in endemic areas (Karl S et al. 2011). The prevalence of SMI, including submicroscopic gametocytes, must be assessed and taken into account in malaria control programs (Okell LC et al. 2009, Karl S et al. 2011).

Only patients with positive blood smears and/or rapid diagnostic tests (RDT) are routinely treated, while the treatment of patients negative by both methods depends on clinical signs and the physician's appreciation. These patients, including those with SMI, may represent more than 10% of infected individuals. In Gabon, SMI currently tends to be more frequent than microscopic infection, possibly due to better preventive policies and/or case management (Bouyou-Akotet et al. 2010). Treatment of all infected subjects, including those with SMI and submicroscopic gametocytes, would reduce the community parasite burden. Indeed, it has been shown that intermittent preventive treatment can reduce the prevalence and genetic diversity of *P. falciparum* malaria (Liljander A et al. 2010).

## ii. *Loa loa*

Human loaiosis differs from other filariasis by the fact that most infected individuals do not have blood microfilariae detectable by standard microscopy. Since the first description of this filariasis, many epidemiologists have found a low prevalence of microfilaria despite local vector abundance. The notion that most patients clear their microfilaremia but continue to have (occult) infection is primarily based on the observation of adult worms during eye passage. The assumption that endemic resistant subjects also may exist (subjects able to completely eliminate *L. loa* infection) is still based on the same observations. Only a sensitive diagnostic test can confirm these assumptions. Our results have shown that 15r3-PCR is suitable for discriminating among endemic groups (microfilaremics, occult infected individuals (occults) and resistant subjects), as the results should be positive in the first two groups and negative in the last. Indeed, two-thirds of infected individuals in southeastern Gabon have occult loaiosis (OL) Touré et al. (1998, 1999a). This needs to be shown in a longitudinal study, however, as *L. loa* infection is characterized by its relative stability in humans and mandrills, the adult worm having a lifespan of about 15 years (Gentillini 1982, Pinder 1994).

This implies that the prevalence of loaiosis would be underestimated by microscopy. If *L. loa* DNA detection is a marker of active infection, all subjects positive by PCR should be treated. This would not have a major impact on health at the individual level but could reduce the parasite burden in the community, in turn reducing the intensity of transmission and resulting in public health benefits.

Studies of resistant individuals may provide interesting immunological information. Marked differences in humoral and cellular immune responses have already been noted between microfilaremic and amicrofilaremic patients (Pinder 1988, Akué 1997, Baize et al. 1997), as well as in the mandrill model (Leroy 1997). However, lacking a reliable method for diagnosing occult infection, it is not known if this difference is due to immunity directed against adult worms or against microfilaria. The identification of endemic groups ("microfilaremics", "occults" and "resistants") by 15r3-PCR method should allow immunological studies to be carried out with sera and cells from each endemic group, using antigens of each developmental stage of *L. loa*, and particularly infective larvae and adult worms. Such studies could help to identify possible cellular or humoral markers involved in resistance to infection, as well as the underlying mechanisms, including host genetic factors. These studies would open the way to investigations of the underlying molecular mechanisms.

In addition, the detection of OL by PCR will allow precise evaluation of filaricide effectiveness during mass treatment, and also that of new drugs in animal models. Pinder et al. showed in 1994 that experimental mandrill infection (*Mandrillus sphinx*) by human *L. loa* isolates led to the same parasitologic characteristics as the natural human infection. Thus, mandrills with occult infection (absence of microfilarae but presence of adult worms, as shown by 15r3-PCR positivity; Touré et al. 1998) can be used to evaluate macrofilaricidal drugs. It has been demonstrated that the 15-kDa polyprotein is conserved within human and simian *L. loa* (Touré et al. 1999b). L.15r3-PCR also detects simian occult *L. loa* and could be used to identify infected animals before their inclusion in preclinical trials.

Finally, serological tests using purified recombinant antigens or peptides offer much better specificity than those using crude antigens. When these antigens become available for loaiosis, immunoenzymatic methods like IgG4 ELISA will reach acceptable specificity. Comparison of ELISA and PCR results sould show whether or not specific IgG4 antibodies are markers of active *L. loa* infection.

## 5. Conclusion

The global strategy of eliminating the parasitic diseases especially malaria and filariasis is mainly based on prevention, early diagnosis and prompt treatment. However, most decisions still rely on microscopy diagnosis which is not always adapted in detecting all infections. Indeed, the success of any intervention depends of the effectiveness of tools and methods especially those allowing proper detection of parasites. PCR offers an exciting opportunity to diagnose submicroscopic malaria infections and occult loaiosis which may constitute a hidden reservoir of disease transmission. The detection of such infections would allow the accurate management of all cases necessary to progress from disease control to elimination.

## 6. References

[1] Levine ND. et al. A newly revised classification of the protozoa (1980). *Protozool.* 27, 37-58.

[2] Levine ND. Protozoan parasites of nonhuman primates as zoonotic agents (1970). *Lab Anim. Care* 20,377-82.

[3] Figtree M, Lee R, Bain L, Kennedy T, Mackertich S, Urban M, Cheng Q, Hudson BJ. Plasmodium knowlesi in human, Indonesian Borneo (2010). *Emerg Infect Dis.* Apr;16 (4):672-4.

[4] Megumi Y L, Alida Coppi, Georges Snounou, Photini Sinnis. Plasmodium sporozoites trickle out of the injection site. *Cell Microbiol.* 1 May 2007;9(5):1215-1222.

[5] Ashley M. Vaughan, Ahmed S. I. Aly, Stefan H. I. Kappe. Malaria parasite pre-erythrocytic stage infection: Gliding and Hiding. *Cell Host Microbe.* 11 September 2008;4(3):209-218.

[6] Olivier Silvie, Maria M Mota, Kai Matuschewski, Miguel Prudêncio. Interactions of the malaria parasite and its mammalian host. *Current Opinion in Microbiology* 2008;11:352-359.

[7] Jake Baum, Dave Richard, Julie Heale et al. A Conserved Molecular Motor Drives Cell Invasion and Gliding Motility across Malaria Life Cycle Stages and Other Apicomplexan Parasites. *The Journal of Biological Chemistry*. February 2006;281:5197-5208.

[8] Sylvia Münter, Benedikt Sabass, Christine Selhuber-Unke et al. Plasmodium Sporozoite Motility Is Modulated by the Turnover of Discrete Adhesion Sites *Cell Host & Microbe*. December 2009;6 (17):551-562.

[9] Malcolm K Jones, Michael F Good. Malaria parasites up close. *Nature Medicine* 2006; 12:170-171.

[10] Kebaier C, Voza T, Vanderberg J. Kinetics of Mosquito-Injected Plasmodium Sporozoites in Mice: Fewer Sporozoites Are Injected into Sporozoite-Immunized Mice. *PLoS Pathog* 2009; 5 (4):e1000399.

[11] Gentillini M., Duflo B. (1982). Filarioses. Médecine Tropicale- Paris: Flammarion. 173-195.

[12] Mogin Observations sur un ver trouvé dans la conjonctive à maribou. île Saint Domingue (1770). Jour. Med. Chirurgie Pharm. etc (Paris). 32 : 338-339. Guyon J (1964). Sur un nouveau cas de filaire, sous conjonctive, ou Filaria occuli des auteurs, observe au Gabon (Côte occidentale d'Afrique). *C. R. Séances Acad. Sci.* 59 : 743-748.

[13] Guyon J. (1864). Sur un nouveau cas de filaire, sous conjonctive, ou Filaria occuli des auteurs, observe au Gabon (Côte occidentale d'frique). *C. R. Séances Acad. Sci.* 59 : 743-748.

[14] Brumpt E. (1904). Les filarioses humaines en Afrique. *Compte Rendu de la Société de Biologie*, 56 : 758-760.

[15] Connor DH, Neafie RC, Meyers WM. (1976). Loiasis : Pathology of tropical and extra-ordinary diseases. Washington: Armed Forces Institute of Pathology. 356-359.

[16] World Health Organization (2009). WHO World Malaria report 2009.

[17] World Health Organization. Severe falciparum malaria (2000).*Trans R Soc Trop Med Hyg*; 94 (Suppl. 1): S1–S90.

[18] Imbert P and Gendrel D (2002). [Malaria treatment in children. 2. Severe malaria]. *Med Trop (Mars)* 62(6): 657-64.

[19] Marsh K, Marsh VM, Brown J, Whittle HC, Greenwood BM (1988). Plasmodium falciparum: the behavior of clinical isolates in an in vitro model of infected red blood cell sequestration.*Exp Parasitol* 65(2):202-8.

[20] Taylor HM, Triglia T, Thompson J, Sajid M, Fowler R, Wickham ME, Cowman AF, Holder AA. (2001). Plasmodium falciparum homologue of the genes for Plasmodium vivax and Plasmodium yoelii adhesrive proteins, which is transcribed but not translated. *Infect Immun*. Jun; 69 (6): 3635-45.

[21] Mazier D, Nitcheu J and Idrissa-Boubou M. (2000). Cerebral malaria and immunogenetics. *Parasite Immunol* 22(12): 613-23.

[22] Brown HC, Chau TT, Mai NT, Day NP, Sinh DX, White NJ, Hien TT, Farrar J, Turner GD. (2000). Blood-brain barrier function in cerebral malaria and CNS infections in Vietnam. *Neurology*. Jul 12;55(1):104-11.

[23] Hunt NH, Grau GE. (2003). Cytokines: accelerators and brakes in the pathogenesis of cerebral malaria. *Trends Immunol*; 24:491-499.

[24] Pino P, Vouldoukis I, Kolb JP, Mahmoudi N, Desportes-Livage I, Bricaire F, Danis M, Dugas B, Mazier D. (2003). Plasmodium falciparum--infected erythrocyte adhesion induces caspase activation and apoptosis in human endothelial cells. *J Infect Dis*. Apr 15;187 (8):1283-90. Epub 2003 Apr 2.

[25] Touré FS, Ouwe-Missi-Oukem-Boyer O, Bisvigou U, Moussa O, Rogier C, Pino P, Mazier D, Bisser S. (2008). Apoptosis: a potential triggering mechanism of neurological manifestation in Plasmodium falciparum malaria. *Parasite Immunol*. Jan;30 (1):47-51.

[26] Combes V, Coltel N, Faille D, Wassmer SC, Grau GE. (2006). Cerebral malaria: role of microparticles and platelets in alterations of the blood-brain barrier. *Int J Parasitol*. May 1;36(5):541-6.

[27] Bisser S, Ouwe-Missi-Oukem-Boyer ON, Toure FS, Taoufiq Z, Bouteille B, Buguet A, Mazier D. (2006). Harbouring in the brain: A focus on immune evasion mechanisms and their deleterious effects in malaria and human African trypanosomiasis. *Int J Parasitol*. May 1;36(5):529-40.

[28] Siau A, Toure FS, Ouwe-Missi-Oukem-Boyer O, Ciceron L, Mahmoudi N, Vaquero C, Froissard P, Bisvigou U, Bisser S, Coppee JY, Bischoff E, David PH, Mazier D. (2007). Whole-transcriptome analysis of Plasmodium falciparum field isolates: identification of new pathogenicity factors. *J Infect Dis*. Dec 1;196(11):1603-12.

[29] Schofield FD, (1955). Two cases of loiasis with peripheral nerve involvement. *Trans. Roy. Soc. Trop. Med. Hyg*. 49 : 588-589.

[30] Hulin C, et al., (1994). Atteinte pulmonaire d'évolution favorable au cours d'une filariose de type *Loa loa. Bull. Soc. Path. Ex*. 87: 248-250.

[31] Okell LC, Ghani AC, Lyons E, Drakeley CJ. (2009). Submicroscopic infection in Plasmodium falciparum-endemic populations: a systematic review and meta-analysis. *J Infect Dis*. Nov 15;200 (10):1509-17.

[32] Snounou, G. S., S. Viriyakosol, X. P. Zhu, W. Jarra, L. Pinheiro, V. E. Do Rosario, S. Thaithong, and K. N. Brown (1993). High sensitivity of detection of human malaria parasites by the use of nested polymerase chain reaction. *Mol. Biochem. Parasitol*. 61:315-320.

[33] Mixson-Hayden T, Lucchi NW, Udhayakumar V. (2010). Evaluation of three PCR-based diagnostic assays for detecting mixed Plasmodium infection. *BMC Res Notes*. Mar 31;3:88.

[34] Jarra W, Snounou G. (1998). Only viable parasites are detected by PCR following clearance of rodent malarial infections by drug treatment or immune responses. *Infect Immun*. Aug;66(8):3783-7.

[35] Cheng Q, Lawrence G, Reed C, Stowers A, Ranford-Cartwright L, Creasey A, Carter R, Saul A. Measurement of Plasmodium falciparum growth rates in vivo: a test of malaria vaccines. *Am J Trop Med Hyg*. 1997 Oct;57(4):495-500.

[36] Filisetti D, Bombard S, N'Guiri C, Dahan R, Molet B, Abou-Bacar A, Hansmann Y, Christmann D, Candolfi E. Prospective assessment of a new polymerase chain reaction target (STEVOR) for imported Plasmodium falciparum malaria. *Eur J Clin Microbiol Infect Dis*. 2002 Sep;21(9):679-81. Epub 2002 Sep 6.

[37] Oyedeji SI, Awobode HO, Monday GC, Kendjo E, Kremsner PG, Kun JF. (2007) Comparison of PCR-based detection of Plasmodium falciparum infections based on single and multicopy genes. *Malar J*. Aug 16;6:112.

[38] Farcas GA, Zhong KJ, Mazzulli T, Kain KC. (2004). Evaluation of the RealArt Malaria LC real-time PCR assay for malaria diagnosis. *J Clin Microbiol*. Feb;42(2):636-8.

[39] Lee MA, Tan CH, Aw LT, Tang CS, Singh M, (2002) Lee SH, Chia HP, Yap EP. Real-time fluorescence-based PCR for detection of malaria parasites. *J Clin Microbiol*. Nov;40(11):4343-5.

[40] Touré FS, et al., (1997a). Species-specific sequence in the repeat 3 region of the gene encoding a putative *Loa loa* allergen: a diagnostic tool for occult loiasis. *Am J Trop Med Hyg*. 56: 57-60.

[41] Touré FS, Odile Bain, Eric Nerrienet, et al. (1997b). Detection of *Loa loa*-specific DNA in blood from occult infected individuals. *Exp Parasitol*. 86: 163-70.

[42] Ouwe-Missi-Oukem-Boyer ON, Hamidou AA, Sidikou F, Garba A, Louboutin-Croc JP. (2005). The use of dried blood spots for HIV-antibody testing in Sahel]. *Bull Soc Pathol Exot*. Dec;98(5):343-6. French.

[43] Plowe CV, Djimde A, Bouare M, Doumbo O, Wellems TE (1995). Pyrimethamine and proguanil resistance-conferring mutations in Plasmodium falciparum dihydrofolate reductase: polymerase chain reaction methods for surveillance in Africa. *Am J Trop Med Hyg* 52: 565–568.

[44] Snounou G, Viriyakosol S, Zhu XP, Jarra W, Pinheiro L, do Rosario VE, Thaithong S, Brown KN. (1993). High sensitivity of detection of human malaria parasites by the use of nested polymerase chain reaction. *Mol Biochem Parasitol*. Oct;61(2):315-20.

[45] Mockenhaupt FP, Ulmen U, von Gaertner C, Bedu-Addo G, Bienzle U.(2002). Diagnosis of placental malaria. *J Clin Microbiol*. Jan;40(1):306-8.

[46] Touré FS, Mezui-Me-Ndong J, Ouwe-Missi-Oukem-Boyer O, Ollomo B, Mazier D, Bisser S. (2006). Submicroscopic Plasmodium falciparum infections before and after sulfadoxine-pyrimethamine and artesunate association treatment in Dienga, Southeastern Gabon. *Clin Med Res*. Sep;4(3):175-9.

[47] Rogier C, Commenges D, Trape JF (1996). Evidence for an age-dependent pyrogenic threshold of Plasmodium falciparum parasitemia in highly endemic populations. *Am J Trop Med Hyg*. Jun;54(6):613-9.

[48] Giha HA, A-Elbasit IE, A-Elgadir TM, Adam I, Berzins K, Elghazali G, Elbashir MI. (2005). Cerebral malaria is frequently associated with latent parasitemia among the semi-immune population of eastern Sudan. *Microbes Infect*. Aug-Sep;7(11-12):1196-203.

[49] Bouyou-Akotet MK, Nzenze-Afene S, Ngoungou EB, Kendjo E, Owono-Medang M, Lekana-Douki JB, Obono-Obiang G, Mounanga M, Kombila M. (2010). Burden of malaria during pregnancy at the time of IPTp/SP implementation in Gabon. *Am J Trop Med Hyg*. Feb;82(2):202-9.

[50] Wongsrichanalai C, Pickard AL, Wernsdorfer WH, Meshnick SR. (2002). Epidemiology of drug-resistant malaria. *Lancet Infect Dis*. Apr; 2 (4):209-18.

[51] Duraisingh MT, Drakeley CJ, Muller O, Bailey R, Snounou G, Targett GA, Greenwood BM, Warhurst DC (1997). Evidence for selection for the tyrosine-86 allele of the pfmdr 1 gene of Plasmodium falciparum by chloroquine and amodiaquine. *Parasitology*. Mar;114 ( Pt 3):205-11.

[52] Coleman RE, Kumpitak C, Ponlawat A, Maneechai N, Phunkitchar V, Rachapaew N, Zollner G, Sattabongkot J. (2004). Infectivity of asymptomatic Plasmodium-infected human populations to Anopheles dirus mosquitoes in western Thailand. *J Med Entomol.* Mar;41(2):201-8.

[53] Nwakanma D, Kheir A, Sowa M, Dunyo S, Jawara M, Pinder M, Milligan P, Walliker D, Babiker HA (2008). High gametocyte complexity and mosquito infectivity of Plasmodium falciparum in the Gambia. *Int J Parasitol.* Feb;38(2):219-27. Epub 2007 Jul 18.

[54] Karl S, Gurarie D, Zimmerman PA, King CH, St Pierre TG, Davis TM. (2011). A submicroscopic gametocyte reservoir can sustain malaria transmission *PLoS One*; 6 (6):e20805. Epub 2011 Jun 14.

[55] Liljander A, Chandramohan D, Kweku M, Olsson D, Montgomery SM, Greenwood B, Färnert A (2010). Influences of intermittent preventive treatment and persistent multiclonal Plasmodium falciparum infections on clinical malaria risk. *PLoS One.* Oct 27;5(10):e13649.

[56] Touré FS, Mavoungou E, Kassambara L, Williams T, Wahl G, Millet P, Egwang TG (1998). Human occult loiasis: field evaluation of a nested polymerase chain reaction assay for the detection of occult infection. *Trop Med Int Health.* 3: 505-511.

[57] Touré FS, Deloron P, Egwang TG, Wahl G (1999a). [Relationship between the intensity of *Loa loa* filariasis transmission and prevalence of infections] *Med Trop.* 59: 249-52. French.

[58] Touré FS, Ungeheuer MN, Egwang TG, Deloron P. (1999b). Use of polymerase chain reaction for accurate follow-up of *Loa loa* experimental infection in *Mandrillus sphinx.* *Am J Trop Med Hyg.* 61: 956-9.

[59] Pinder M, Everaere S, Roelants GE (1994). Loa loa: immunological responses during experimental infections in mandrills (Mandrillus sphinx). *Exp Parasitol.* Sep; 79(2):126-36.

[60] Pinder M Dupont A, Egwang TG. (1988). Identification of a surface antigen on *Loa loa* microfilariae the recognition of which correlates with the amicrofilaremic state in man. *Jour. Immunol.* 141 : 2480-2486.

[61] Akué JP, Hommel M., Devaney E. (1997). High levels of parasite- specific IgG1 correlate with the amicrofilaremic state in *Loa loa* infection. *Jour. Infect. Dis.* 175 : 158-163.

[62] Baize S, Wahl G, Soboslay PT, Egwang TG, Georges AJ. (1997). T helper responsiveness in human *Loa loa* infection: defective specific proliferation and cytokine production by CD4+ T cells from microfilaremic subjects compared with amicrofilaremics. *Clin. Exp. Immunol.* 108 : 272-278.

[63] Leroy E, Baize S, Wahl G, Egwang TG, Georges AJ. (1997). Experimental infection of a nonhuman primate with Loa loa induces transient strong immune activation followed by peripheral unresponsiveness of helper T cells. *Infection Immunity.* 65 (5) : 1876-1882.

# 10

# Detection of Bacterial Pathogens in River Water Using Multiplex-PCR

C. N. Wose Kinge[1], M. Mbewe[2] and N. P. Sithebe[1]
*[1]Department of Biological Sciences, School of Environmental and Health Sciences,*
*North-West University, Mafikeng Campus, Mmabatho,*
*[2]Animal Health Programme, School of Agricultural Sciences, North-West University,*
*Mafikeng Campus, Mmabatho*
*South Africa*

## 1. Introduction

The aquatic environments receive a significant number of human microbial pathogens from point and non-point sources of pollution. Point-source pollution enters the environment at different locations, through a direct route of discharge of treated or untreated domestic sewage, industrial effluent and acid mine drainage (State of the Environment Report [SER], 2002). Non-point (or diffuse) sources of pollution comprises up to 80 % of the pollution entering major river systems thus are of significant concern with respect to the dissemination of pathogens and their indicators in water systems. They may be attributable to the run-off from urban and agricultural areas, leakage from sewers and septic systems, insecticides and herbicides from agricultural land, and sewer overflows (Stewart et al., 2008). Although majority of pathogenic microbes can be eliminated by sewage treatment, many end up in the effluent which is then discharged into receiving bodies of water. These pathogenic microbes have been implicated in human diseases linked with the use of contaminated water and food. Adequate sanitation and clean water, being two critical factors in ensuring human health, protects against a wide range of water-related diseases. These include diarrhoea, cholera, trachoma, dysentery, typhoid, hepatitis, polio, malaria, and filariasis (United Nations Department of Public Information [UNDPI], 2005).

Water is a vital natural resource because of its basic role to life, quality of life, the environment, food production, hygiene, industry, and power generation (Meays et al., 2004). With the rapid increase in world population and increased urbanisation, there is a massive strain on the existing water supply and sanitation facilities (UNDPI, 2005). In the developing world, poor access to safe water and inadequate sanitation continues to be a danger to human health (World Health Organisation [WHO], 2004). The water situation, in the African continent, has attracted a lot of concern from all sectors of government as it is estimated that more than 300 million out of the 800 million people who live on the continent are in water-scarce environments (United Nations Educational, Scientific and Cultural Organisation [UNESCO], 2004). In Northern Africa, the present water supply is unstable as population growth and economic development have surpassed the traditional water management practices, leading to water scarcity and pollution to a varying degree (UNESCO, 2004). According to Beukman and Uitenweerde (2002), Southern Africa faces very serious water challenges with an estimated half of the population lacking access to portable water and sanitation facilities. They further hinted that, by 2025, countries like Mozambique, Namibia, Tanzania and Zimbabwe will face more water pressures.

The scarcity of water does not only threaten food security, but also the production of energy and environmental integrity. This often results in water usage conflicts between different communities, and water contamination when humans and animals share the same source of water (Kusiluka et al., 2005). According to the Department of Water Affairs – DWA (2000), South Africa is a water scarce region, with 450mm rainfall per annum. This is lower than the world's 860mm average rainfall. Of the forty-four million people who live in South Africa, 12 million people were without access to portable water supply prior to 1994 (Momba et al., 2006). Although the South African government is making significant progress in ensuring the supply of potable water to all communities, 3.3 and 15.3 million inhabitants of South Africa are still identified to be living without access to potable water and adequate sanitation facilities (Council for Scientific and Industrial Research [CSIR], 2008). A total of 80% of the population live in the rural areas with the unavailability of potable basic water supplies and proper sanitation facilities (Kasrils, 2004; Reitveld et al., 2009).

Due to the scarcity of water in South Africa, extensive exploitation of water resources such as dams, pools, unprotected rivers and springs for domestic and other water uses, is common, particularly in the rural communities where access to potable water supply is limited (Younes and Bartram, 2001). In many developing countries with inadequate sanitation, faecal contaminations of environmental waters by enteric pathogens are very common and river water is major source of microbial pathogens (Sharma et al., 2010). In this study, we report the use of conventional identification, and multiplex PCR (m-PCR) method that permits the simultaneous detection of water-borne *Salmonella, Shigella, E. coli,* and *Klebsiella* bacteria spp. from rivers in the North West province of South Africa. The major rivers in the province include the Molopo, Groot Marico, Elands, Hex, and Crocodile Vaal, Skoonspruit, Harts and Mooi. These rivers are grouped into five catchment areas, which include the Crocodile and Elands, Marico and Hex, Marico and Molopo, Mooi and Vaal, and the Harts (SER, 2002; Department of Water Affairs [DWA], 2007). The water quality in these rivers has been impaired partly due to the frequent contamination of water sources with a number of pathogenic microorganisms from human as well as animal activities, which result in the spread of diarrhoeal diseases (Meays et al., 2004).

## 1.1 Bacterial pathogens in the aquatic environment

Microbial pathogens in water include viruses, bacteria, and protozoa (Girones et al., 2010). Currently, pathogenic bacteria have been identified as the major etiological agent in the majority of the waterborne outbreaks worldwide (WHO 2003; Liang et al., 2006). Bacillary dysentery caused by *Shigella* bacteria alone is responsible for approximately 165 million cases of bacterial diarrhoeal diseases annually. Of this, 163 million are in developing countries and 1.5 million in industrialized ones accounting for an estimated 1.1 million death cases each year (Sharma et al., 2010). Most members of the genus *Arcobacter* have been isolated from different environmental water sources including surface and ground water. Their presence has been correlated with that of faecal pollution indicators (Collado et al., 2008; Fong et al., 2007; Ho et al., 2006) as well as meat mainly from poultry, pork and beef (Collada et al., 2009; Houf, 2010; Wesley and Miller, 2010). Some members of the genus *Arcobacter*, like *A. butzleri*, *A. cryaerophilus*, and *A. skirrowii*, have been implicated in animal and human diarrhoeal cases, suggesting a faecal oral route of transmission to humans and animals (Gonzalez et al., 2007). *Helicobacter pylori* on the other hand, found to be present in surface water and wastewater has been implicated in gastritic, peptic, and duodenal ulcer diseases (Linke et al., 2010; Queralt et al., 2005).

Biofilms in drinking water distribution systems have been reported as possible reservoirs of H. *pylori* and attempts to culture these cells from water samples have proven unsuccessful (Linke et al., 2010; Percival and Thomas, 2009). Due to the fastidious nature of this bacterium, the lack of standard culture methods for environmental samples, and the controversy in its ability to survive in an infectious state in the environment, very few quantitative studies have been reported (Percival and Thomas, 2009). *Legionella pneumophila* is a ubiquitous bacterium in natural aquatic environments that can also persist in human-controlled systems containing water, such as air conditioning and plumbing infrastructures (Steinert et al., 2002). Furthermore, *Vibrio vulnificus*, an opportunistic human pathogen that cause gastroenteritis, severe necrotizing soft-tissue infections and primary septicaemia, has been found present in fish, shell fish, water, and wastewater. Infection generally, is associated with the ingestion of contaminated seafood and water (Harwood et al., 2004; Igbinosa et al., 2009). More so, the presence of enteric bacteria of the genera *Salmonella*, *Shigella*, *E.coli* and *Klebsiella* in water has been identified as a major threat to human health and causative agents for many diseases (Leclerc et al., 2001).

*Salmonellae are* the most frequent agents of bacterial gastroenteritis and typhoid in humans and a prime example of a water- and shell fish-transmitted human pathogen. It is frequently isolated from the marine environment where it can remain viable for several hours (Malorny et al., 2008; Westrell et al., 2009). Contamination with *Salmonella* has been reported in surface water used for recreational purposes, source of drinking water (Till et al., 2008) and irrigation (Gannon et al. 2004) underlining the possible risk associated to the use of such contaminated water. The typhoid caused by *Salmonella enterica* serotype Typhi remains an important public health problem in developing countries and the burden of typhoid fever worldwide is further compounded by the spread of multiple drug resistant *S. typhi* (Kim 2010; Lynch et al., 2009; Srikantiah et al., 2006). The runoff from fields with animal husbandry, and untreated sewage disposal contribute to the presence of *Salmonella* in natural water resources (Jenkins et al., 2008; Moganedi et al., 2007). Low numbers of *Salmonella* in food, recreational, surface and potable water sources may pose a public health risk given that their infective dose can be as low as 15–100 CFU (Cobbold et al., 2006; Seo et al., 2006).

Species of *Shigella* and enteroinvasive *Escherichia coli* (EIEC) are important human pathogens identified as the major cause of bacillary dysentery (Wanger et al., 1988; Szakál et al., 2003). The infective dose of *Shigella* cells is very low ($10^1$-$10^4$ organisms), whereas EIEC strains require a larger infectious dose (between $10^6$ and $10^{10}$ organisms) (Rowe and Gross, 1984). Both *Shigella* spp. and EIEC carry a large invasion plasmid and express a similar set of proteins. Both of them are transmitted by direct contact from human to human or via contaminated food and water (Parsot, 1994; Rowe and Gross, 1984). Clinical features of bacillary dysentery caused by EIEC that resemble shigellosis include fever, severe abdominal cramps, malaise, toxemia, and watery diarrhea. The serotype *E. coli* O157:H7, an emergent pathogen of faecal origin frequently isolated from waters, has been implicated in food and water-borne disease outbreaks (Bavaro, 2009).

Bacteria of the genus *Klebsiella* are ubiquitous in nature and are a frequent cause of nosocomial infections (Horan et al., 1988). Their non-clinical habitats encompass the gastrointestinal tract of mammals as well as environmental sources such as soil, surface waters, and plants (Bagley, 1985). Environmental isolates have been described as being indistinguishable from human clinical isolates with respect to their biochemical reactions and virulence (Matsen et al., 1974). While the medical significance of *Klebsiella* obtained in the natural environment is far from clear, such habitats are thought to be potential reservoirs for the growth and spread of these bacteria which may colonize animals and humans (Knittel et al., 1977). Of the five identified *Klebsiella* species, *K. oxytoca* and *K. Pneumonia*, remain the most clinically important opportunistic pathogen, implicated in community-acquired pyogenic liver abscess and bacterial meningitis in adults (Casolari et al., 2005; Haryani et al., 2007; Keynan and Rubinstein, 2007), has been reported to be present in water (Syposs et al., 2005).

## 1.2 Methods used in detection of bacterial pathogens in water

Detection, differentiation, and identification of bacteria can be performed by numerous methods, including phenotypic, biochemical and immunological assays, and molecular techniques. These traditional methods for the detection and enumeration of bacterial pathogens have largely depended on the use of selective culture and standard biochemical methods. This classical microbiological methodology relies on the cultivation of specific bacteria, for example plate counts of coliforms. Drawbacks of these methods include firstly, pathogenic bacteria, which normally occur in low numbers, tend to incur large errors in sampling and enumeration (Fleischer, 1990). Secondly, culture-based methods are time-consuming, tedious; detect only one type of pathogen, and no valid identification of the pathogen (Szewzyk et al., 2000). Thirdly, many pathogenic microbes in the environment, although viable, are either difficult to culture or are non-culturable (Roszak and Colwell, 1987). Sometimes too, it is often difficult to achieve appropriate enrichment, which makes the work even more tedious.

Moreover, concentrations may be too low for cultural detection but still be high enough to cause infection. These limitations therefore make routine examination of water samples for pathogens like *Vibrio cholerae*, *Shigella dysenteriae*, *Aeromonas* spp. and *Campylobacter* spp., difficult. Instead, bacterial indicator species like *Escherichia* coli, which is a normal flora present in very high numbers in the gut of warm-blooded animals, is widely used as an indicator of faecal pollution, to estimate the risk of exposure to other pathogenic microbes present in animal or human wastes (Lund, 1994). However, *Escherichia coli* as well as some bacterial species like *Enterococcus faecalis*, once released into freshwater bodies, enter a viable but non-culturable (VBNC) state and express different set of activities, including virulence traits (Lleo et al., 2005). As a result, the current methodology is unsuitable for the detection of bacterial pathogens in water and the assessment of their virulence potential. Therefore, a molecular detection method is needed, since such methods are highly specific and sensitive.

Molecular methods used are typically based on the detection and quantification of specific segments of the pathogen's genome (DNA or RNA). To achieve this, the specific segments are subjected to *in vitro* amplification. These methods allow researchers to speedily and specifically detect microorganisms of public health concern (Girones et al., 2010). Recently, molecular techniques, specifically nucleic acid amplification procedures, immunocapturing, fluorescence *in-situ* hybridization (FISH), and polymerase chain reaction (PCR) have provided highly sensitive, rapid and quantitative analytical tools for detecting specific pathogens in environmental samples (Watson et al., 2004). These techniques are used to evaluate the microbiological quality of food and water, as well as microbial source-tracking (Albinana-Gimenez et al., 2009; Field et al., 2003; Hundesa et al., 2006). Most applied molecular techniques are based on protocols of nucleic acid amplification, of which the polymerase chain reaction (PCR) is the most commonly used.

PCR is a molecular tool that allows for the amplification of target DNA fragments using oligonucleotide primers in a chain of replication cycles catalysed by DNA polymerase (*Taq* polymerase) (Rompré et al., 2002). This tool is used for microbial identification and surveillance with high sensitivity and specificity (Watterworth et al., 2005). It has successfully been applied for the detection and identification of pathogenic bacteria in clinical and environmental samples, as well as for the investigation of food and water-borne disease outbreaks (Harakeh et al., 2006; Haryani et al., 2007; Hsu and Tsen, 2001; Riyaz-Ul-Hassan et al., 2004; Shabarinath et al., 2007). The use of quantitative PCR (qPCR) is rapidly becoming established in the environmental sector since it has been shown, in many cases, to be more sensitive than either the bacterial culture method or the viral plaque assay (He and Jiang, 2005). However, molecular protocols, unlike traditional culture-based methods, do not distinguish between viable and non-viable organisms hence the need for more information before replacing the current conventional methods by molecular ones.

Molecular techniques for the specific detection and quantification of bacterial pathogens also offer several advantages over conventional methods: high sensitivity and specificity, speed, ease of standardization and automation. As with the viruses, direct PCR amplification of some bacterial pathogens from water samples is difficult due to the presence of only low numbers of these bacteria in environmental sources. Therefore, an enrichment step is usually required prior to performing a PCR (Noble and Weisberg, 2005). Improved detection of pathogenic *E. coli* (Ogunjimi and Choudary, 1999) by immuno-capture PCR, and the sensitive detection of *Salmonella* (Hoorfar et al., 2000) and *Campylobacter* (Nogva et al., 2000) by real-time PCR have also been developed; but these procedures are all mono-specific and are either laborious or very expensive for routine use in water testing laboratories. More recent improvements have allowed simultaneous detection of several microorganisms in a single assay (Maynard et al., 2005; Straub et al., 2005; Marcelino et al., 2006). The use of such multiplex polymerase chain reaction (m-PCR) has provided rapid and highly sensitive methods for the specific detection of pathogenic microbes in the aquatic environment (Kong et al., 2002).

## 1.3 Multiplex polymerase chain reaction (m-PCR)

Following the application of PCR in the simultaneous amplification of multiple loci in the human dystrophin gene (Chamberlain et al., 1998), multiplex PCR has been firmly established as a general technique. To date, the application of multiplex PCR in pathogen identification, gender screening, linkage analysis, template quantitation, and genetic disease diagnosis is widely established (Chehab and Wall, 1992; Kong et al., 2002; Serre et al., 1991; Shuber et al., 1993). For pathogen identification, PCR analysis of bacteria is advantageous, as the culturing and typing of some pathogens has proven difficult or lengthy. Bacterial multiplexes indicate a particular pathogen among others, or distinguish species or strains of the same genus. An amplicon of sequence conserved among several groups is often included in the reaction to indicate the presence of phylogenetically or epidemiologically similar, or environmentally associated, bacteria and to signal a functioning PCR. Multiplex assays of this set-up distinguish species of Legionella (Bej et al., 1990), Mycobacterium (Wilton and Cousin, 1992), Salmonella (Chamberlain et al., 1998), Escherichia coli, and Shigella (Bej et al., 1991) and major groups of Chlamydia (Kaltenboek et al., 1992) from other genus members or associated bacteria. It has also been shown that multiplex PCR remains the ideal technique for DNA typing because the probability of identical alleles in two individuals decreases with the number of polymorphic loci examined. Reactions have been developed with potential applications in paternity testing, forensic identification, and population genetics (Edwards et al., 1991, 1992; Klimpton et al., 1993). Multiplex PCR can be a two-amplicon system or it can amplify 13 or more separate regions of DNA. It may be the end point of analysis, or preliminary to further analyses such as sequencing or hybridization. The steps for developing a multiplex PCR and the benefits of having multiple fragments amplified simultaneously, however, are similar in each system (Edwards and Gibbs, 1994).

### 1.4 Aim/Objectives of the study

To detect the presence of pathogenic Escherichia coli, Klebsiella, Salmonella, and Shigella species in water samples obtained from rivers in the North-West Province of South Africa, conventional typing and multiplex PCR methods were applied to enriched cultures. The objectives of the study were to use conventional methods to check for the presence and molecular tools to confirm the identity of Escherichia, Klebsiella, Salmonella, and Shigella species in river water. Our prognosis is that the results will emphasize the need for a rapid and accurate detection method for water-borne disease outbreaks and bacterial pathogens in water to protect human health.

## 2. Materials and methods

A total of 54 water samples were collected using sterile 500mL McCartney bottles, downstream, midstream, and upstream of the Crocodile, Elands, Hex, Mooi, Vaal, Molopo, Groot Marico, Harts and Skoonspruit rivers between November 2007 and March 2008 (Fig. 1). These rivers form the five major catchments in the province, which are the Crocodile and Eland, Marico and Hex, Marcio and Molopo, Mooi and Vaal, and Harts catchments. Samples collected were transported on ice to the laboratory for analysis.

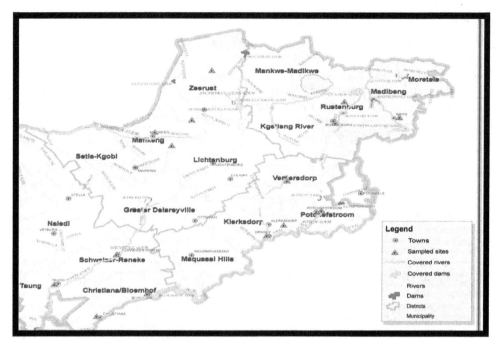

Fig. 1. A cross-section of the North West province map showing the rivers and dams sampled

### 2.1 Bacterial reference strains

Bacterial strains used for the experimental work (Table 1) were American Type Culture Collection (ATCC) cultures. The strains were grown on Nutrient Agar (Biolab, Merck, South Africa) under aerobic conditions at 37°C for 24 hours.

| Bacterial Strains | Source | Reference | *Mdh* | *IpaH* | *IpaB* | *GapA* |
|---|---|---|---|---|---|---|
| *Salmonella paratyphi* | ATCC 9150 | This study | – | – | + | – |
| *Salmonella typhi* | ATCC 14028 | Hsu and Tsen, 2001 | – | – | + | – |
| *Shigella boydii* | ATCC 9207 | Wang et al., 1997 | – | + | – | – |
| *Shigella sonnei* | ATCC 25931 | Wang et al., 1997 | – | + | – | – |
| *Klebsiella pneumonia* | ATCC 15611 | Lu et al., 2000 | – | – | – | + |
| *K. oxytoca* | ATCC 43086 | This study | – | – | – | + |
| *Escherichia coli* | ATCC 25922 | Lu et al., 2000 | – | – | – | – |

Table 1. Bacterial strains used in the study for evaluation of primer specificity

## 2.2 Selective isolation of *Salmonella*, *Shigella*, *E. coli* and *Klebsiella*

Water analysis for *Salmonella, Shigella, E. coli* and *Klebsiella* bacteria, was done using the spread plate method (American Public Health Association [APHA], 1998). In brief, 1mL of each water sample was enriched in 9mL of 2% buffered-peptone water (Biolab, Merck Diagnostics, South Africa) and serial dilutions performed. Aliquots of 0.1mL of each dilution were plated out on Eosin Methylene Blue (EMB) agar plates (Biolab, Merck Diagnostics, South Africa) for the presumptive isolation of *E. coli* and *Klebsiella*, and on Salmonella-Shigella agar for *Salmonella* and *Shigella* isolation. All plates were incubated at 37°C for 24 hours. Presumptive isolates were sub-cultured on fresh media plates incubated at 37°C for 24hours and then preserved on 2.3% w/v Nutrient agar plates for further analysis.

## 2.3 Bacterial Identification using triple sugar iron (TSI) agar test

All 2992 and 1180 presumptive isolates on EMBA and SSA plates, respectively were Gram stained using the method of Cruikshank et al., (1975) to confirm their morphology as Gram negative rod-shaped bacteria. All Gram negative isolates were subjected to the TSI test, a biochemical test, which distinguishes the *Enterobacteriaceae* from other intestinal Gram-negative bacilli by the ability of the organisms to catabolise the sugars glucose, lactose and sucrose present at different concentrations in the medium, and the production of acid and gas (Prescott, 2002). The test was performed as previously recommended (United States Pharmacopeia Convention; Inc., 2001). Briefly, isolates were streak-plated on TSI agar slopes and incubated at 37°C for 24hours. The results were interpreted as previously determined by Forbes and Weissfeld (1998).

## 2.4 Differentiation of *Salmonella*, *Shigella*, *E. coli* and *Klebsiella* using conventional serological assay

All *Salmonella, Shigella, E. coli* and *Klebsiella* candidate isolates obtained from culture plates and identified by Triple Sugar Iron [TSI] agar test, were differentiated by conventional serotyping (Ballmer et al., 2007). To test for surface antigens, *E. coli* Poly D1–D8; *Shigella boydii* Poly C, C1, C2 and C3, *Shigella dysenteriae* Poly A Types 1, 2, 3, 4, 5, 6, 7, *Shigella sonnie* Poly D Phase I and II, *Shigella flexneri* Poly B Types I, II, III, IV, V, VI; *Salmonella* O Poly O (Factors A–G, O2, O4, O7, O8, O9, O9, 46, O3, 10, O1, 3, 4) and O1 (Factors O11, O13, O6, 14, O16, O18, O21, O35), *Salmonella* H Poly Phase 1 and 2; and *Klebsiella* Capsular Types 1, 2, 3, 4, 5, 6 antisera (Inqaba Biotech, South Africa) were used.

## 2.5 DNA extraction

Genomic DNA was extracted from all positive bacteria cells inoculated in 5mL Luria Bertani (LB) broth (Merck, South Africa) following overnight incubation at 37°C in a shaker (Doyle and Doyle, 1990). The pellets obtained were re-suspended in 50µL of sterile distilled water. The concentration of the extracted DNA in solution was determined spectrophotometrically (UV Visible spectrophotometer model S-22, Boeco, Germany) by measuring the absorbance at 260 nm. The DNA in solution was used as a template for multiplex PCR.

## 2.6 Oligonucleotide primers and multiplex PCR method

Oligonucleotide primers used in the study were synthesized by Inqaba Biotech, South Africa. Sequences of the four PCR primer pairs for m-PCR, their corresponding gene targets and size of the expected amplifications are as shown (Table 2). The malate dehyrogenase gene (*Mdh*) of *E. coli* (Hsu and Tsen, 2001; Wose Kinge and Mbewe, 2011), the invasive pasmid antigen B gene (*IpaB*) of *Salmonella* spp. (Kong et al., 2002), the invasive plasmid antigen gene H (*IpaH*) of *Shigella* spp. (Kong et al., 2002; Wose Kinge and Mbewe, 2010), and the glyceralehye-3-phospahate dehydrogenase gene (*GapA*) genes for *Klebsiella* spp. (Diancourt et al., 2005; Wose Kinge and Mbewe, 2011) were simultaneously detected by multiplex polymerase chain reaction (m-PCR) assays. DNA from 50µL extract from enriched cultures was used for PCR amplification in a final volume of 25µL. The reaction mixture consisted of 2X PCR Master mix (0.05µL Taq DNA polymerase, 4mM MgCl$_2$, 0.4mM dNTPs) (Fermentas, Inqaba Biotechnical Industries (Pty) Ltd, South Africa), 0.3µM of IpaB, 0.2µM of IpaH and 1.0µM each of Mdh and GapA genes. PCR amplification was performed in a Peltier Thermal Cycler (model-PTC-220 DYAD™ DNA ENGINE; MJ Research Inc. USA) under the following conditions: heat denaturation at 94°C for 3 min, followed by 34 cycles of denaturation at 94°C for 30 s; annealing at 60°C for 60 s and extension at 72°C for 1 min. This was followed by a final extension step at 72°C for 7 min and 4°C hold. To create a negative control template DNA was excluded.

| Organism | Target gene | Primer | Primer sequence (5'→3') | Expected size (bp) |
|---|---|---|---|---|
| *E. coli* | *Mdh* | Mdh F | CGTTCTGTTCAAATGGCCTCAGG | 392 |
| | | Mdh R | ACTGAAAGGCAAACAGCCAAG | |
| *Salmonella* | *IpaB* | IpaB F | GGACTTTTTAAAGCGGCGG | 314 |
| | | IpaB R | GCCTCTCCCAGAGCCGTCTGG | |
| *Shigella* | *IpaH* | IpaH F | CCTTGACCGCCTTTCCGATA | 606 |
| | | IpaH R | CAGCCACCCTCTGAGGTACT | |
| *Klebsiella* | *GapA* | GapA F | GTTTTCCCAGTCACGACGTTGTATGAA ATATGACTCCACTCACG | 700 |
| | | GapA R | TTGTGAGCGGATAACAATTTCCTTCAG AAGCGGCTTTGATGGCT | |

Table 2. Oligonucleotide primers used in this study

## 2.7 Electrophoresis and visualization of PCR products

Following amplification, 10µL of each sample was electrophoresed in a horizontal agarose (LONZA, South Africa) 1% w/v slab gel containing ethidium bromide (0.1µg/mL) in 1X TAE buffer (40 mM tris-acetate; 2 mM EDTA, pH8.3). The agarose gel was electrophoresed for six hours at 60 V. The gel was visualized with UV light (Gene Genius Bio Imaging System, SYNGENE model GBOX CHEMI HR). The relative molecular sizes of the PCR products were estimated by comparing their electrophoretic mobility with 100bp marker (Fermentas O' GeneRuler DNA ladder; Canada).

## 2.8 Specificity of primers

The specificity of the primers used for multiplex-PCR was confirmed against related enteric bacterial DNA. The DNA was extracted from 5mL of overnight bacterial suspensions cultured in Luria Bertani broth as described under section 2.5. The extracted DNA was then stored at -20°C for use in m-PCR.

## 3. Results and discussion

### 3.1 Differentiation of *Salmonella*, *Shigella*, *E. coli* and *Klebsiella* using conventional serotyping assay

In order to differentiate the bacterial isolates using surface antigens present, conventional serotyping by slide agglutination was performed using polyvalent antisera. The commercially available typing antisera are not sufficient to recognize all prevalent serotypes of *Salmonella*, *E. coli* and *Klebsiella* spp. In our study, the antisera assay was not used to identify these serotypes, but rather to determine if a given isolate was a member of the genera of interest or not. The percentages of *E. coli*, *Klebsiella*, *Shigella* and *Salmonella* isolates obtained, showing a positive agglutination to antisera, were calculated for each catchment area and results recorded as contained in Table 3. The results indicate a presence of *E. coli*, *Klebsiella*, *Shigella* and *Salmonella* spp. in all five catchments areas. According to Table 3, *E. coli* (the main indicator for faecal contamination) was present in all five catchment samples. The highest was 29% in the Crocodile and Elands catchment, followed by the Mooi and Vaal catchment with 24% agglutination with surface antigen specific antisera. The other three catchments were not free of *E. coli* although at lesser levels, comparably.

According to DWA and WHO standards, water meant for irrigation (DWA, 1996) and human consumption (WHO, 2001) should contain no *E. coli* bacteria. The use of such contaminated water for irrigation as well as direct consumption as it is before treatment would result in the transmission of potentially pathogenic bacteria to humans through contaminated vegetables and other crops eaten raw, as well as milk from grazing cattle. *Klebsiella* was highest in the Mooi and Vaal followed by Harts catchments with 19% and 11%, respectively. Podschun et al. (2001) also reported a high percentage (53%) distribution of *Klebsiella* spp. from surface water samples, the most common species being *K. pneumoniae*. Bacteria species of the genera *Escherichia* and *Klebsiella* are amongst the group of faecal coliforms. Generally, faecal coliform bacteria inhabit the gastrointestinal tract of all warm and some cold-blooded animals as normal commensals, hence their presence in any given water body is a clear indication of faecal contamination. Although their presence in water cannot be pinpointed to a specific source of faecal contamination, faecal material from human and animal sources can be regarded as high risk due to the possible presence of pathogenic bacteria (Harwood et al., 2000).

High levels of *Shigella* contamination were also seen in all catchments with 31% and 41% in the Crocodile and Elands catchment and Harts catchment, respectively. In general, there was lesser contamination with *Salmonella* compared to other faecal coliforms in all catchments with a maximum of 8% in Mooi and Vaal catchment. Water-borne pathogens often occur in reasonably low concentrations in environmental waters. Therefore, some form of filtration and proliferation are needed for pathogen detection (Hsu et al., 2010). Following filtration of the sample on membrane filters, bacteria retained on filters can then be detected by culturing in or on selective media. Additional steps, such as biochemical tests, serological assays, and molecular methods, are necessary for confirmation. The isolation and identification of *Shigella* spp. and *E. coli* are straightforward and well established (Echeverria et al., 1991, 1992). However, *Shigella* spp. and entero-invasive *E. coli* [EIEC] are genetically close and exhibit considerable antigenic cross-reactivity, thus differentiating between them using a single method can be difficult (Cheasty and Rowe, 1983; Lan et al., 2001; Kingombe et al., 2005; Yang et al., 2005).

The O and H antigen serotyping method provide important epidemiological information. However it is not appropriate for routine diagnostic use because of its high cost and the labour-intensive requirements (Ballmer et al., 2007). There is, therefore, an urgent need for an accurate and simple detection, identification, and differentiation technique for *Shigella* spp. and EIEC, especially for epidemiological studies. On the contrary, serotyping is currently the most widely used technique for typing *Klebsiella* species. It is based mainly on a division according to the K (capsule) antigens (Ørskov and Ørskov, 1984) and shows good reproducibility and capability in differentiating most clinical isolates (Ayling-Smith and Pitt, 1990).

| River Catchments | *E. coli* % | *Klebsiella* % | *Shigella* % | *Salmonella* % |
|---|---|---|---|---|
| Crocodile and Elands | 29 | 4 | 37 | 6 |
| Marico and Hex | 9 | 7 | 18 | 4 |
| Marico and Molopo | 9 | 4 | 12 | 1 |
| Mooi and Vaal | 24 | 19 | 15 | 8 |
| Harts | 7 | 11 | 41 | 6 |

Table 3. Prevalence of *E. coli*, *Klebsiella*, *Shigella* and *Salmonella* bacteria obtained by serotyping

### 3.2 Multiplex PCR

The m-PCR was designed to target genes specific to the four entero-pathogenic bacteria selected for this study. Results obtained showed the presence of *E. coli*, *Klebsiella*, *Shigella* and *Salmonella* contamination in the five catchment areas (Table 4). A total of 39% of *E. coli* was recorded for the Crocodile and Elands catchment and up to 45% of *Shigella* spp. was recovered from the Marico and Hex catchment. The presence of *Klebsiella* and *Salmonella* spp. was also observed with 10% and 11% in the Mooi and Vaal catchment, respectively. Of these bacteria species, contamination with *Shigella* was widespread in all catchments. Detection of the *IpaH* gene, which is present on both the chromosome and the *inv* plasmid of all *Shigella* spp., confirmed the presence of this bacterium in water (Hsu and Tsen, 2001). Understanding the ecology of *Shigella* had been limited mainly due to the lack of suitable techniques to detect the presence of *Shigella* in environment samples (Faruque et al., 2002).

In the present study, we used molecular techniques as well as conventional serotyping method to detect *Shigella* as well as *E. coli*, *Salmonella* and *Klebsiella* spp. in river waters with special reference to virulence genes. We standardized the assay by culturing the environmental water samples and simultaneously conducting m-PCR tests. In a similar study by Faruque et al. (2002) and Sharma et al. (2010), the *IpaH* gene was used as an indicator tool to detect the presence of *Shigella* in environmental waters. Fresh contamination of surface water by faecal material of dysentery patients is a possibility in developing countries where sanitation is poor resulting in the presence of *Shigella* in surface water. Several previous studies have also detected *Shigella* in surface waters or sewage samples and have indicated that *Shigella* strains can possibly be transported by surface waters (Alamanos et al., 2000; Faruque et al., 2002; Obi et al., 2004a; Pergram et al., 1998).

Similarly, amplification of the *Mdh* gene, which codes for malic acid dehydrogenase, a housekeeping enzyme of the citric acid cycle, and reportedly found in all *E. coli* strains (Hsu and Tsen, 2001), confirmed the presence of both commensal and pathogenic *E. coli* in the water samples. Although *E. coli* is usually present as harmless commensals of the human and animal intestinal tracts, pathogenic strains possess virulent factors that enable them to cause diseases and hence, constitute a potential risk to the health of consumers (Kuhnert et al., 2000). For the detection of *Salmonella* spp. the *IpaB* gene, which is a virulence gene found on the invasion plasmid of *Salmonella* spp., was selected for the PCR as it is reportedly present in most *Salmonella* strains (Kong et al., 2002). *Salmonella* is isolated from water in lower numbers than indicator bacteria such as faecal coliforms, faecal streptococci and enterococci, which are several orders of magnitude higher (Sidhu and Toze, 2009).

However, low numbers (15-100 colony-forming units [CFU]) of *Salmonella* in water may pose a public health risk (Jyoti et al., 2009). In the aquatic environment this pathogen has been repeatedly detected in various types of natural waters such as rivers, lakes, coastal waters, estuarine as well as contaminated ground water (Haley et al., 2009; Levantesi et al., 2010; Lin and Biyela, 2005; Moganedi et al., 2007; Theron et al., 2001; Wilkes et al., 2009). Their presence has been attributable to runoff from fields with animal husbandry, addition of untreated sewage from nearby civilization contribute *Salmonella* in natural water resources (Moganedi et al., 2007; Jenkins et al., 2008). *Salmonella* contaminated waters might contribute through direct ingestion of the water or via indirect contamination of fresh food to the transmission of this microorganism. *Salmonella* prevalence in surface water and drinking water has not been uniformly investigated in different countries in recent papers.

Surveys of *Salmonella* in fresh surface water environment were mainly performed in industrialized nations, particularly in Canada and North America. Reports of *Salmonella* prevalence in drinking water were instead more frequent from developing nations reflecting the higher concern relating to the use of low quality drinking water in these countries. Overall, the scientific community has mainly recently focused on the prevalence of is microorganism in impacted and non-impacted watersheds (Haley et al., 2009; Jokinen et al., 2011; Patchanee et al., 2010), on the identification of the routes of salmonellae contamination (Gorski et al., 2011; Jokinen et al., 2010, 2011; Obi et al., 2004b; Patchanee et al., 2010), and on the influence of environmental factors on the spread of *Salmonella* in water (Haley et al., 2009; Jokinen et al., 2010; Meinersmann et al., 2008; Wilkes et al., 2009).

Although direct consumption of water by humans from these rivers was minimal throughout the study, indirect consumption through fishing was common. This was particularly evident in the Crocodile and Elands, Marico and Molopo, and the Mooi and Vaal catchment areas. This may be a cause for concern because fish in water bodies contaminated with human and animal waste, harbour a considerable number of bacteria such as *Salmonella*, *Clostridium botulinum*, *Vibrio cholerae*, *E. coli* and other coliforms, which could be transmitted to humans if eaten raw or under-cooked (Jayasinghe and Rajakaruna, 2005). Fish and shellfish accounts for 5% of individual cases and 10% of all food-borne illness outbreaks in the United States (Flick, 2008) and not only does fish constitute potential sources of bacteria, they also harbour antibiotic resistant bacteria that could be transmitted to humans resulting in the spread of a pool of antibiotic resistant genes into the environment (Miranda and Zemelman, 2001; Pathak and Gopal, 2005). This also might be compounded by the presence of opportunistic pathogens like *Klebsiella* species in water with serious health implications for consumers that utilize water directly or indirectly from the rivers, especially high risk patients with impaired immune systems such as the elderly or young, patients with burns or excessive wounds, those undergoing immunosuppressive therapy or those with HIV/AIDS infection. Colonization may lead to invasive infections and on very rare occasions, *Klebsiella* spp., notably *K. pneumoniae* and *K. oxytoca*, may cause serious infections, such as destructive pneumonia (Bartram et al., 2003; Genthe and Steyn, 2006).

| River Catchments | E. coli % | Klebsiella % | Shigella % | Salmonella % |
|---|---|---|---|---|
| Crocodile and Elands | 39 | 0 | 11 | 6 |
| Marico and Hex | 4 | 6 | 45 | 0 |
| Marico and Molopo | 0 | 6 | 5 | 1 |
| Mooi and Vaal | 15 | 10 | 5 | 11 |
| Harts | 0 | 0 | 23 | 9 |

Table 4. Prevalence of *E. coli*, *Klebsiella*, *Shigella* and *Salmonella* bacteria obtained by m-PCR

### 3.3 Specificity of primers

In order to evaluate and verify the specificity of the primers in this study, each primer pair was tested by PCR on DNA templates prepared from a panel of seven different bacterial control strains. The analysis indicated that all primer pairs showed specificities only for their corresponding target organisms (Table 1) and all four sets of PCR primers were targeted at a virulence-associated gene. The Mdh primers specifically amplified a 392bp malic acid dehydrogenase gene fragment from *E. coli* strain obtained from the American Type Culture Collection (Table 1) and 4-39% of isolates obtained from the different river catchments. The IpaH primers produced a specific 606bp amplimer in all *Shigella* spp. examined in this study (Table 1; Fig 2. lane 3), which included two species of the genus, viz., *S. sonnei* and *S. boydii*, which are known to be pathogenic to humans. In a previously reported study, Kong et al.

(2002) tested two virulence genes of *Shigella*, the *virA* gene and the *IpaH* gene and obtained more positive amplifications with the *IpaH* gene when compared with the *virA* gene.

Although the *virA* gene was previously reported by Villalobo and Torres (1998) to be specific for virulent *Shigella* spp., the *IpaH* gene was found to be more reliable in detecting *Shigella* spp. in environmental isolates (Kong et al., 2002; Wose Kinge and Mbewe, 2010). The IpaB primers were found to produce a specific 314bp amplimer, in all *Salmonella* spp. examined, which included *S. paratyphi*, and *S. typhimurium* (Table 1; Fig 2. Lanes 7 and 8) as well as 1-11% of the isolates tested. Similar results were obtained with the GapA primers which generated a 700bp amplimer specific to *Klebsiella*. The amplimers were confirmed by sequencing (Inqaba Biotech, South Africa) all showed a high percentage of sequence similarity (>90%) with published malic acid dehydrogenase, invasive plasmid antigen H and B, and glyceraldehydes-3-phosphate gene sequences in the GenBank database. Our results therefore, indicated that this particular set of primers were suitable for the specific detection of most general strains of *E. coli*, *Salmonella*, *Shigella* and *Klebsiella* from water samples.

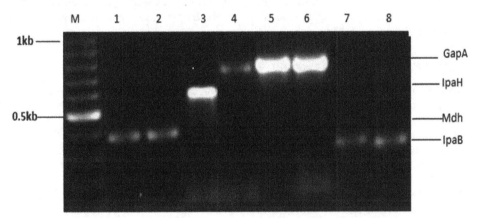

Fig. 2. Electrophoretic analysis of PCR-amplified target genes from six different bacterial pathogens. Mobilities of the different target gene amplicons are indicated on the right. Lane M, 100bp DNA ladder (size marker); lanes 1 and 2, Mdh amplicon of *Escherichia coli* ATCC 25922; lane 3, IpaH amplicon of *Shigella boydii* ATCC 9207; lane 4, GapA amplicon of *Klebsiella oxytoca* ATCC 43086; lanes 5 and 6, GapA amplicon of *K. pneumoniae* ATCC 15611; lane 7, IpaB amplicon of *Salmonella paratyphi* ATCC 9150, lane 8, IpaB amplicon of *S. typhimurium* ATCC 14028

## 4. Conclusion

Both conventional and molecular methods successfully identified bacteria of interest, however, the multiplex-PCR assays were sensitive and faster than conventional serotyping methods for detecting *E. coli*, *Salmonella*, *Shigella*, and *Klebsiella* spp. from river water samples. The 392bp *Mdh*, 314bp *IpaB*, 606bp *IpaH* and 700bp *GapA* genes were found to be specific and present in the control strains analyzed. Therefore, m-PCR screening of these strains for *Mdh*, *IpaB*, *IpaH* and *GapA* genes should provide a better indicator of possible presence of potentially pathogenic *E. coli, Salmonella, Shigella* and *Klebsiella* bacteria in river water. The water quality is affected by human activities around the areas, which include industrial processes, mining, agriculture and domestic usage. Thus, the main source of *E. coli, Salmonella, Shigella* and *Klebsiella* in these rivers may be discharge from wastewater effluent as well as domestic sewage around the catchment areas. Our results indicate that the water-borne and food-borne spread of these pathogens is possible due to drinking water contamination, recreational activities, and fisheries. Since the aquatic environment is implicated as the reservoir for these microorganisms, and consequently responsible for their transmission in humans, it is obvious that detailed studies on the pathogenic potential of the environmental strains will certainly contribute to understanding the virulence properties of these bacteria and to establish the importance of these significant pathogens of aquatic systems. The results thus emphasize the need for the implementation of a rapid and accurate detection method in cases of water-borne disease outbreaks and the need for more rapid detection of bacterial pathogens in water to protect human health. The ability to rapidly monitor for various types of microbial pathogens would be extremely useful not only for routine assessment of water quality to protect public health, but also allow effective assessments of water treatment processes to be made by permitting pre- and post-treatment waters to be rapidly analyzed.

## 5. References

Alamanos, Y., Maipa, V., Levidiotou, S., Gessouli, E. (2000). A community waterborne outbreak of gastro-enteritis attributed to *Shigella sonnei. Epidemiol. Infect.* 125, pp. (499–503)

Albinana-Gimenez, N., Miagostovich, M., Calgua, B., Huguet, J.M., Matia, L., & Girones, R. (2009). Analysis of adenoviruses and polyomaviruses quantified by qPCR as indicators of water quality in source and drinking water-treatment plants. *Water Res.* 43, pp. (2011-2019)

APHA. (1998). Standard methods for the examination of Water and Wastewater, 19th Ed. Amer. Public Health Assoc. Washington DC.

Ayling-Smith, B., & Pitt, TL. (1990). State of the art in typing: *Klebsiella* species. *J. Hosp. Infect.* 16, pp. (287–295)

Bagley, ST. (1985). Habitat association of *Klebsiella* species. *Infect. Control*, 6, pp. (52–58)

Ballmer, K., Korczak, BM., Kuhnert, P., Slicker, P., Ehricht, R., & Hächler, H. (2007). Fast DNA serotyping of *Escherichia coli* by use of an oligonucleotide microassay. *J. Clin. Microbiol.* 45, 2, pp. (370-379)

Bartram, J., Chartier, Y., Lee, JV., Pond, K., & Surman-Lee S (Eds) (2007): *Legionella and the Prevention of Legionellosis.* WHO Press, World Health Organization, Geneva, Switzerland.

Bartram, J., Cotruvo, J., Exner, M., Fricker, C., & Glasmacher, A. (2003). Heterotrophic plate counts in drinking water safety: the significance of HPCs for water quality and human health. WHO Emerging Issues in Water and Infectious Disease Series. London, IWA Publishing.

Bavaro, MF. (2009). *Escherichia coli* O157: what every internist and gastroenterologist should know. *Curr. Gastroenterol. Rep.* 11, 4, pp. (301-306)

Bej, AK., Mahbubani, MH., Miller, R., DiCesare, JL., Haft, L., & Atlas, RM. (1990). Multiplex PCR amplification and immobilized capture probes for detection of bacterial pathogens and indicators in water. *Mol. Cell. Prob.* 4, pp. (353-365).

Bej, AK., McCarty, SC., & Atlas, RM. (1991). Detection of coliform bacteria and *Escherichia coli* by multiplex polymerase chain reaction: Comparison with defined substrate and plating methods for water quality monitoring. *Appl. Environ. Microbiol.* 57, pp. (1473-1479)

Beukman, R. & Uitenweerde, T. (2002). Water Demand Management Phase II. International Conservation Union, Pretoria, pp. (12)

Casolari, C.., Pecorari, M., Fabio, G., Cattani, S., Venturelli, C., Piccinini, L., Tamassia, M.G., Gennari, W., Sabbatini, A.M., Leporati, G., Marchegiano, P., Rumpianesi, F., & Ferrari, F. (2005). A simultaneous outbreak of *Serratia marcescens* and *K. pneumoniae* in a neonatal intensive care unit. *The J. Hosp. Infec.* 61, pp. (312-320)

Chamberlain, JS., Gibbs, RA., Ranier, JE., Nguyen, PN., & Caskey, CT. (1988). Deletion screening of the Duchenne muscular dystrophy locus via multiplex DNA amplification. *Nucleic Acids Res.* 16, pp. (11141-11156).

Cheasty, T., & Rowe, B. (1983). Antigenic relationships between the enteroinvasive *Escherichia coli* O antigens O28ac, O112ac, O124, O136, O143, O144, O152, and O164 and *Shigella* O antigens. *J. Clin. Microbiol.* 17, pp. (681–684)

Chehab, FF., & Wall, J. (1992). Detection of multiple cystic fibrosis mutations by reverse dot blot hybridization: A technology for carrier screening. *Hum. Genet.* 89, pp. (163-168)

Cobbold, RN., Rice, DH., Davis, MA., Besser, TE., & Hancock, DD. (2006). Long-term persistence of multidrug-resistant *Salmonella enterica* serovar Newport in two dairy herds. *J. Amer. Vet. Med. Assoc.* 228, pp. (585–91)

Collado, L., Guarro, J., & Figueras, MJ. (2009) Prevalence of *Arcobacter* in meat and shellfish. *J. Food. Prot.* 72, pp. (1102– 1106)

Collado, L., Inza, I., Guarro, J., & Figueras, M.J. (2008) Presence of *Arcobacter* spp. in environmental waters correlates with high levels of fecal pollution. *Environ. Microbiol.* 10, pp. (1635–1640)

Collie, T.K., Strom, M.S., Sinigalliano, C.D., Moeller, P.D., & Holland, A.F. (2008). The coastal environment and human health: microbial indicators, pathogens, sentinels and reservoirs. *Environ. Health,* 7, 2, pp. (S3)

Cruikshank, R., Duguid, JP., Marmoin, BP.,& Swain, RH. (1975). Medical Microbiology, 12th ed. Longman, New York. 2, pp. (3-4)

CSIR. (2008). Water Quality and Pollution: Parliamentary Portfolio Committee on Water Affairs and Forestry.

Diancourt, L., Passet, V., Verhoef, J., Grimont, P.A., & Brisse, S. (2005). Multilocus sequence typing of *Klebsiella pneumoniae* nosocomial isolates. *J. Clin. Microbiol.* 42, 8, pp. (4178-4182)

Doyle, JJ., & Doyle, JL. (1990). A rapid total DNA preparation procedure for fresh plant tissue. *Focus,* 12, pp. (13-15).

DWA. (2000). Water use authorisation process for individual applications, DWAF, Edition 1 final draft for implementation and use, revision 3, December 2000.

DWA. (2007). Integrated Vaal River system WMR studies. Department of Water Affairs and Forestry, Pretoria.

Echeverria, P., Sethabutr, O., & Pitarangsi, C. (1991). Microbiology and diagnosis of infections with *Shigella* and enteroinvasive *Escherichia coli*. *Rev. Infect. Dis.* 13, Suppl. 4, pp. (S220–S225)

Echeverria, P., Sethabutr, O., Serichantalergs, O., Lexomboon, U., & Tamura, K. (1992). *Shigella* and enteroinvasive *Escherichia coli* infections in households of children with dysentery in Bangkok. *J. Infect. Dis.* 165, pp. (144–147)

Edwards, A., Civitello, A., Hammond, HA., & Caskey, CT. (1991). DNA typing and genetic mapping with trimeric and tetrameric tandem repeats. *Am. J. Hum. Genet.* 49, pp. (746–756)

Edwards, A., Hammond, HA., Jin, L., Caskey, CT., & Chakroborty, R. (1992). Genetic variation at five trimeric and tetrameric tandem repeat loci in four human population groups. *Genomics*, 12, pp. (241-253)

Edwards, MC., & Gibbs, RA. (1994). Multiplex PCR: advantages, development and applications. *Genome Res.* 3, pp. (S65- S75)

Faruque, SM., Khan, R., Kamruzzaman, M., Yamasaki, S., Ahmad, QS., Azim, T., Nair, GB., Takeda, Y., & Sack, DA. (2002). Isolation of *Shigella dysenteriae* type 1 and *S. flexneri* strains from surface waters in Bangladesh: comparative molecular analysis of environmental *Shigella* isolates versus clinical strains. *Appl. Environ. Microbiol.* 68, pp. (3908–3913)

Field, KG., Bernhard, AE., & Brodeur, TJ. (2003). Molecular approaches to microbiological monitoring: faecal source detection. *Environ. Monit. Assess.* 81, pp. (313-326)

Fleisher, JM. (1990). Conducting recreational water quality surveys: some problems and suggested remedies. *Mar Pollut Bull.* 21, 12, pp. (562–7)

Flick, GJ. (2008). Microbiological safety of farmed fish. Global Aquaculture Advocate, pp. (33-34)

Fong, TT., Mansfield, LS., Wilson, DL., Schwab, DJ., Molloy, SL., & Rose, JB. (2007). Massive microbiological groundwater contamination associated with a waterborne outbreak in Lake Erie, South Bass Island, Ohio. *Environ. Health Perspect.* 115, 6, pp. (856-864)

Forbes, S., & Weissfeld, F. (1998). Bailey and Scott's diagnostic microbiology, 10th Ed. Mosby, Inc., St. Louis, Mo. Harakeh, S., Yassine, H., El-Faidel, M. (2006). Antibiotic resistant patterns of *Escherichia coli* and *Salmonella* strains in the aquatic Lebanese environments. *Environ. Pollut.* 143, pp. (269-277)

Gannon, VP., Graham, T., Read, S., Ziebell, K., Muckle, A., & Mori, J., et al. (2004). Bacterial pathogens in rural water supplies in Southern Alberta, Canada. *J. Toxicol. Environ. Health*, 67, 20–22, pp. (1643–1653)

Genthe, B., & Steyn, M. (2006). Good intersectoral water governance- a Southern African decision-makers guide. Chapter on Health and Water. CSIR/NRE/WR/EXP/2006/0047/A.

Girones, R., Ferrus, MA., Alonso, JL., Rodriguez-Manzano, J., Correa AA., Calgua, B., Hundesa, A., Carratala, A., & Bofill- Mas, S. (2010). Molecular detection of pathogens in water- The pros and cons of molecular techniques. *Water Res.* 44, pp. (4325-4339).

González, A., Botella, S., Montes, RM., Moreno, Y., & Ferrús, M.A. (2007). Direct detection and identification of *Arcobacter* species by multiplex PCR in chicken and wastewater samples from Spain. *J. Food Prot.* 70, (341-347)

Gorski, L., Parker, CT.,Liang, A., Cooley, MB., Jay-Russell, MT., Gordus, AG. (2011). Prevalence, distribution and diversity of *Salmonella enterica* in a major produce region of California. *Appl. Environ. Microbiol.* doi: 10.1128/AEM.02321-10.

Haley, B.J., Cole, DJ., & Lipp, EK. (2009). Distribution, diversity and seasonality of water-borne *Salmonella* in a rural watershed. *Appl. Environ. Microbiol.* 75, pp. (1248–1255)

Haley, BJ., Cole, DJ., & Lipp, EK. (2009). Distribution, diversity and seasonality of water-borne *Salmonella* in a rural watershed. *Appl. Environ. Microbiol.* 75, pp. (1248–1255)

Harakeh, S., Yassine, H., & EL- Fadel, M. (2006). Antimicrobial resistant patterns of *Escherichia coli* and *Salmonella* strains in the aquatic Lebanese environments. *Environ. Pol.* 143, 2, pp. (269-277)

Harwood, VJ., Gandhi, JP., & Wright, AC. (2004). Methods for isolation and confirmation of *Vibrio vulnificus* from oysters and environmental sources: a review. *J. Microbiol. Methods,* 59, pp. (301-316)

Harwood, VJ., Whitlock, J., & Withington, V. (2000). Classification of antibiotic resistance patterns of indicator bacteria by discriminant analysis: use in predicting the sources of faecal contamination in subtropical waters. *Appl. Environ. Microbiol.* 66, 9, pp. (3698-3704)

Haryani, Y., Noorzaleha, A.S., Fatimah, A.B., Noorjahan, B.A., Patrick, G.B., Shamsinar, A.T., Laila, R.A.S., & Son, R. (2007). Incidence of *Klebsiella pneumoniae* in street foods sold in Malaysia and their characterization by antibiotic resistance, plasmid profiling, and RAPD-PCR analysis. *Food Cont.* 18, pp. (847-853)

He, J., & Jiang, S. (2005). Quantification of enterococci and human adenoviruses in environmental samples by real-time PCR. *Appl. Environ. Microbiol.* 71, 5, pp. (2250-2255)

Ho, HT., Lipman, LJ., & Gaastra, W. (2006) *Arcobacter,* what is known and unknown about a potential food-borne zoonotic agent!. *Vet. Microbiol.* 115, pp. (1–13)

Hoorfar, J., Ahrens, P., & Radstrom, P. (2000). Automated 5'Nuclease PCR Assay for Identification of *Salmonella enterica. J Clin Microbiol.* 38, 9, pp. (3429–35)

Horan, T., Culver, D., Jarvis, W., Emori, G., Banerjee, S., Martone, W., & Thornsberry, C. (1988). Pathogens causing nosocomial infections. *Antimicrob. Newsl.* 5, pp. (65–67)

Houf, K. (2010) *Arcobacter.* In: Dongyou, L. (Ed.), Molecular detection of foodborne pathogens, CRC Press, New York, USA, pp. (289–306)

Hsu, B-M., Wu, S-F., Huang, S-W., Tseng, Y-J., Chen, J-S., Shih, F-C., & Ji, D-D. (2010). Differentiation and identification of *Shigella* spp. and enteroinvasive *Escherichia coli* in environmental waters by a molecular method and biochemical test. *Water Res.* 44, pp. (949-955)

Hsu, S-C., & Tsen, H-Y. (2001). PCR primers designed from malic acid dehydrogenase gene and their use for detection of *Escherichia coli* in water and milk samples. *Int. J. Food Microbiol.* 64, pp. (1-11)

Hundesa, A., Maluquer de Motes, C., Bofill-Mas, S., AlbinanaGimenez, N., & Girones, R. (2006). Identification of human and animal adenoviruses and polyomaviruses for determination of sources of fecal contamination in the environment. *Appl. Environ. Microbiol.* 72, pp. (7886-7893)

Igbinosa, EO., Obi, LC., & Okoh, AI. (2009). Occurrence of potentially pathogenic vibrios infinal effluents of a wastewater treatment facility in a rural community of the Eastern Cape Province of South Africa. *Res. Microbiol.* 160, 8, pp. (531-537)

Jayasinghe, PS., & Rajakuruna, RMAGG. (2005). Short Communication: Bacterial contamination of fish sold in fish markets in the Central province of Sri Lanka. *J. Nat. Sci. Foundation Sri Lanka*, 33, 3, pp. (219-221)

Jenkins, MB., Endale, DM., & Fisher, DS. (2008) Most probable number methodology for quantifying dilute concentrations and fluxes of *Salmonella* in surface waters. *J Appl Microbiol*, 104, pp. (1562-8)

Jokinen, C., Edge, TA., Ho, S., Koning, W., Laing, C., Mauro, W., et al. (2011). Molecular subtypes of Campylobacter spp., *Salmonella enterica*, and *Escherichia coli* O157:H7 isolated from fecal and surface water samples in the Oldman River watershed, Alberta, Canada. Water Res. 45, pp. (1247-1257)

Jyoti, A., Ram, S., Vajpayee, P., Singh, G., Dwivedi, PD., Jain, SK., & Shanker, R. (2009). Contamination of surface and potable water in South Asia by Salmonellae: Culture-independent quantification with molecular beacon real-time PCR. *Sci. Total Environ.* 408, 6, pp. (1256-1263)

Kaltenboek, B., K.G. Kansoulas, and J. Storz. 1992. Two-step polymerase chain reactions and restriction endonuclease analyses detect and differentiate *ompA* DNA of the *Chlamydia* spp. *I. Clin. Microbiol.* 30, pp. (1098-1104)

Kasrils, R. (2004). A decade of delivery. Minister of Water Affairs and Forestry.

Keynan, Y., & Ruinstein, E., 2007. The emerging face of *K. pneumoniae* infections in the community. *Int. J. Antimicrob. Agents.* 30, pp. (385-389)

Kim, HM., Hwang, CY., & Cho, BC. (2010). *Arcobacter marinus* sp. nov. *Int. J. Syst. Evol. Microbiol.* 60, pp. (531–536)

Kingombe, CI., Cerqueira-Campos, M-L., & Farber, JM. (2005). Molecular strategies for the detection, identification and differentiation between enteroinvasive *Escherichia coli* and *Shigella* spp. *J. Food Prot.* 68, 2, pp. (239–245)

Klimpton, CP., Gill, P., Walton, A., Urquhart, A., Millican, E.S., & Adams, M. (1993). Automated DNA profiling employing multiplex amplification of short tandem repeat loci. *PCR Methods Applic.* 3, pp. (13-21)

Knittel, MD., Seidler, RJ., Eby, C., & Cabe, LM. (1977). Colonization of the botanical environment by *Klebsiella* isolates of pathogenic origin. *Appl. Environ. Microbiol.* 34, pp. (557–563)

Kong, RYC., Lee, SKY., Law, TWF., Law, SHW., & Wu, RSS. (2002). Rapid detection of six types of bacteria pathogens in marine waters by multiplex PCR. *Water Res.* 36, pp. (2802-2812)

Kuhnert, P., Boerlin, P., & Frey, J. (2000). Target genes of virulence assessment of *Escherichia coli* isolates from water, food and environment. *Fed. European Microbiol. Soc. Rev.* 24, pp. (107-117)

Kusiluka, LJM., Karimuribo, ED., Mdegela, RH., Luoga, EJ., Munishi, PKT., Mlozi, MRS., & Kambarage, DM. (2005). Prevalence and impact of water-borne zoonotic pathogens in water, cattle and humans in selected villages in Dodoma Rural and Bagamoyo districts, Tanzania. *Phy. Chem. Earth*, 30, pp. (818-825)

Lan, R., Lumb, B., Ryan, D., Reeves, PR. (2001). Molecular evolution of large virulence plasmid in *Shigella* clones and enteroinvasive *Escherichia coli*. *Infect. Immun.* 69, pp. (6303–6309)

LeClerc, H., Mossel, DAA., Edberg, SC., & Struijk, CB. (2001). Advances in the bacteriology of the coliform group: Their suitability as markers of microbial water safety. *Ann. Rev. Microbiol.* 55, (201-234)

Levantesi, C., LaMantia, R., Masciopinto, C., Böckelmann, U., Ayuso-Gabella, MN., & Salgot, M. (2010). Quantification of pathogenic microorganisms and microbial indicators in three wastewater reclamation and managed aquifer recharge facilities in Europe. *The Sci. Total Environ.* 408, 21, pp. (4923–4930)

Liang, JL., Dziuban, EJ., Craun, GF., Hill, V., Moore, MR., Gelting, RJ., Calderon, RL., Beach, MJ., & Roy, SL. 2006. Surveillance for waterborne disease and outbreaks associated with drinking water and water not intended for drinking — United States, 2003–2004. MMWR Surveillance Summit 2006, 55, pp. (5512-5561)

Lin, J., & Biyela, PT. (2005). Convergent acquisition of antibiotic resistance determinants amongst the *Enterobacteriaceae* isolates of the Mhlathuze River, KwaZulu-Natal (RSA). *Water SA.* 31, 2, pp. (0378-4738), ISSN

Linke, S., Lenz, J., Gemein, S., Exner, M., & Gebel, J. (2010). Detection of *Helicobacter pylori* in biofilms by real-time PCR. *Int. J. Hyg. Environ. Health,* 213, pp. (176-182)

Lleo, MM., Bonato, B., Benedetti, D., & Canepari, P. (2005). Survival of enterococcal species in aquatic environments. *FEMS Microbiol. Ecol.* 54, pp. (189-196)

Lund, V. (1994). Evaluation of *Escherichia coli* as an indicator for the presence of *Campylobacter jejuni* and *Yersinia* enterocoliticain chlorinated and untreated oligotrophic lake water. Abstract HRM37, Water Quality International 94. IAWQ 17 Biennial International Conference, Budapest, Hungary, 24–29 July 1994.

Lynch, MF., Blanton, EM., Bulens, S., Polyak, C., Vojdani, J., & Stevenson, J. (2009). Typhoid fever in the United States, 1999–2006. *J. Amer. Med. Assoc.* 302, 8, pp. (852–869)

Malorny, B., Löfström, C., Wagner, M., Krämer, N., & Hoorfar, J. (2008). Enumeration of *Salmonella* bacteria in food and feed samples by real-time PCR for quantitative microbial risk assessment. *Appl Environ Microbiol,* 74, pp. (1299– 304)

Marcelino, L.A., Backman, V., Donaldson, A., Steadman, C., Thompson, J.R., Preheim, S.P., Lien, C., Lim, E., & Veneziano, D. (2006). Accurately quantifying low-abundant targets amid similar sequences by revealing hidden correlations in oligonucleotide microarray data. *Polz MF. Proc. Natl. Acad. Sci. U. S. A.* 103, 37, pp. (13629-13634)

Matsen, JM., Spindler, JA., & Blosser, RO. (1974). Characterization of *Klebsiella* isolates from natural receiving waters and comparison with human isolates. *Appl. Microbiol.* 28, pp. (672–678)

Maynard, C., Berthiaume, F., Lemarchand, K., Harel, J., Payment, P., Bayardelle, P., Masson, L., & Brousseau, R. (2005). Waterborne pathogen detection by use of oligonucleotide-based microarrays. *Appl. Environ. Microbiol.* 71, 12, pp. (8548-8557)

Meays, C., Broersma, K., Nordin, R., & Mazumder, A. (2004). Source tracking faecal bacteria critical review of current methods. *J. Environ. Man.* 73, pp. (71-79)

Meinersmann, RJ., Berrang, ME., Jackson, CR., Fedorka-Cray, P., Ladely, S., & Little, E. (2008). *Salmonella, Campylobacter* and *Enterococcus* spp.: Their antimicrobial resistance profiles and their spatial relationships in a synoptic study of the Upper Oconee River basin. Microbial Ecology, 55, 3, pp. (444–452)

Miranda, CD., & Zemelman, R. (2001). Antibiotic resistant bacteria in fish from concepción bay, Chile. *Marine Pol. Bulletin,* 42, 11, pp. (1096-1102)

Moganedi, KLM., Goyvaerts, EMA., Venter, SN., & Sibara, MM. (2007). Optimisation of the PCR-*invA* primers for the detection of *Salmonella* in drinking and surface waters following a pre-cultivative step. *Water SA*. 33, 2, pp. (195-201)

Momba, MNB., Tyafa, Z., Makala, N., Brouekaert, BM., & Obi, CL. (2006). Safe drinking water is still a dream in the rural areas of South Africa. Case study: The Eastern Cape Province. *Water SA*. 32, 5, pp. (715-720)

Noble, RT., & Weisberg, SB. (2005). A review of technologies for rapid detection of bacteria in recreational waters. *J. Water Health*, 3, 4, pp. (381-392)

Nogva, HK., Bergh, A., Holck, A., & Rudi, K. (2000). Application of the 50-nuclease PCR assay in evaluation and   development of methods for quantitative detection of *Campylobacter jejuni*. *Appl Environ Microbiol*. 6, 9, pp. (4029– 36)

Obi, CL., Bessong, PO., Momba, MNB., Potgieter, N., Samie, A., & Igumbor, E.O. (2004a). Profiles of antibiotic   susceptibilities of bacterial isolates and physico-chemical quality of water supply in rural Venda communities,  South Africa. *Water SA*. 30, 4, pp. (515-520)

Obi, CL., Potgieter, N., Musie, EM., Igumbor, EO., Bessong, PO., Samie, A. (2004b). Human and environmental-associated   non-typhoidal *Salmonella* isolates from different sources in the Venda region of South Africa. Proceedings of the   2004 Water Institute of Southern Africa (WISA), Biennial Conference, May 2004.

Ogunjimi, AA., & Choudary, PV. (1999). Adsorption of endogenous polyphenols relieves the inhibition by fruit juices and   fresh produce of immuno-PCR detection of *Escherichia coli* O157: H7. *FEMS Immunol Med Microbiol*. 23, 3, pp.  (213–20)

Ørskov, I., & Ørskov, F. (1984). Serotyping of *Klebsiella*. *Methods Microbiol*. 14, pp. (143–164)

Parsot, C. (1994). *Shigella flexneri*: genetics of entry and intercellular dissemination in epithelial cells. In: Dangl, J.L. (Ed.), Bacterial Pathogenesis of Plants and Animals. *Curr. Top. Microbiol. Immunol*. 192, pp. (217–241)

Pathak, SP., & Gopal, K. (2005). Occurrence of antibiotic and metal resistance in bacteria from organs of river fish. *Environ. Res*. 98, pp. (100-103)

Percival, SL., & Thomas, JG. (2009). Transmission of *Helicobacter pylori* and the role of water and biofilms. *J. Water Health*, 7, 3, pp. (469-477)

Pergram, GC., Rollins, N., & Esprey, Q. (1998). Estimating the costs of diarrhoea and epidemic dysentery in KwaZulu- Natal and South Africa. *Water SA*. 24, 1, pp. (11–20)

Podschun, R., Pietsch, S., Holler, C., & Ullmann, U. (2001). Incidence of *Klebsiella* species in surface waters and their   expression of virulence factors. *Appl. Environ. Microbiol*. 67, 7, pp. (3325-3327)

Prescott, H. (2002). Laboratory exercise in Microbiology. Smith publishers, New York. 5th Ed. pp. (133)

Queralt, N., Bartolomé , R., & Araujo, R. (2005). Detection of *Helicobacter pylori* DNA in human faeces and water with different levels of faecal pollution in the north-east of Spain. *J. Appl. Microbiol*. 98, pp. (889-895)

Reitveld, LC., Haarhoff, J., & Jagals, P. (2009). A tool for technical assessment of rural water supply systems in South Africa. *Phy. Chem. Earth*, 34, pp. (43-49)

Riyaz-Ul-Hassan, S., Verma, V., & Qazi, G.N. (2004). Rapid detection of *Salmonella* by polymerase chain reaction. *Molecul. Cellul. Probes*, 18, pp. (333-339)

Rompré, A., Servais, P., Baudart, J., de-Roubin, M-R., & Laurent, P. (2002). Detection and enumeration of coliforms in drinking water: current methods and emerging approaches. *J. Microbiol. Methods* 49, pp. (31-34)

Roszak, DB., & Colwell, RR. (1987). Survival strategies of bacteria in the natural environment. *Microbiol. Rev.* 51, pp. (365– 79)

Rowe, B., & Gross, RJ. (1984). Facultatively anaerobic Gram negative rods. Genus II. *Shigella.* In: Krieg, NR., Holt, JG. (Eds.), Bergey's Manual of Systematic Bacteriology. Williams & Wilkins, Baltimore, MD, pp. (423–427)

Seo, KH., Valentin-Bon, IE., & Brackett, RE. (2002). Detection and enumeration of *Salmonella enteritidis* in homemade ice cream associated with an outbreak: comparison of conventional and real-time PCR methods. *J Food Prot.* 69, pp. (639–43)

SER. (2002). State of Emergency Report, North West Province, South Africa.

Serre, JL., Taillandier, A., Mornet, E., Simon-Bouy, B., Boue, J., & Boue, A. (1991). Nearly 80% of cystic fibrosis heterozygotes and 64% of couples at risk may be detected through a unique screening of four mutations by ASO reverse dot blot. *Genomics* 11, pp. (1149-1151)

Shabarinath, S., Sanath Kumar, H., Khushiramani, R., Karunasagar, I., & Karunasagar, I. (2007). Detection and characterization of *Salmonella* associated with tropical seafood. *Int. J. Food Microbiol.* 114, pp. (227-233).

Sharma, A., Singh, SK., & Bajpai, D. (2010). Phenotypic and genotypic characterization of *Shigella* spp. with reference to its virulence genes and antibiogram analysis from river Narmada. *Microbiol. Res.* 165, pp. (33-42)

Shuber, AP., Skoletsky, J., Stern, R., & Handelin, BL. (1993). Efficient 12-mutation testing in the CFTR gene: A general model for complex mutation analysis. *Hum. Mol. Genet.* 2, pp. (153-158)

Sidhu, JPS., & Toze, SG. (2009). Human pathogens and their indicators in biosolids: a literature review. *Environ. Int.* 35, pp. (187-201)

Srikantiah, P., Girgis, FY., Luby, SP., Jennings, G., Wasfy, MO., & Crump, JA. (2006). Population-based surveillance of typhoid fever in Egypt. *The Amer. J. Trop. Med. Hyg.* 74, pp. (114–119)

Steinert, M., Hentschel, U., & Hacker, J. (2002). *Legionella pneumophila*: an aquatic microbe goes astray. *FEMS Microbiol. Rev.* 26, pp. (149-162).

Stewart, J.R., Gast, R.J., Fujioka, R.S., Solo-Gabriele, H.M., Meschke, J.S., Amaral-Zettler, L.A., Del Castillo, E., Polz, M.F.,

Straub, TM., Dockendorff, BP., Quinonez-Diaz, MD., Valdez, CO., Shutthanandan, JI., Tarasevich, BJ., Grate, JW., & BrucknerLea, CJ. (2005). Automated methods for multiplexed pathogen detection. *J. Microbiol. Methods,* 62 (3), pp. (303-316)

Syposs, Z., Reichart, O., & Meszaros, L., 2005. Microbiological risk assessment in the beverage industry. *Food Control*, 16, 6, pp. (515-521)

Székal, D., Schneider, G., & Pál, T. (2003). A colony blot immune assay to identify enteroinvasive *Escherichia coli* and *Shigella* in stool samples. *Diagn. Microbiol. Infect. Dis.* 45, 3, pp. (165–171)

Szewzyk, U., Szewzyk, R., Manz, W., & Schleifer, H. (2000). Microbial safety of drinking water. *Annu. Rev. Microbiol.* 54, (81-127)

Theron, J., Morar, D., Preez, MDU., Brozel, VS., & Venter, SN. (2001). A sensitive semi-nested PCR method for the detection of *Shigella* in spiked environmental water samples. *Water Res.* 35, pp. (869–874)

Till, D., McBride, G., Ball, A., Taylor, K., & Pyle, E. (2008). Large-scale freshwater microbiological study: Rationale, results and risks. *J. Water and Health*, 6, 4, pp. (443–460)

UNDPI. (2005). The International Decade for Action: "Water for Life" 2005-2015. United Nations Department of Public Information.

UNESCO. (2004). Water Programme for Africa, Arid and Water Scarce Zones 2004-2006.

United States Pharmacopeia Convention, Inc. (2001). The United States Pharmacopeia 25. Rockville, M.D.

Villalobo, E., & Torres A. (1998). PCR for detection of *Shigella* spp. in mayonnaise. *Appl. Environ. Microbiol.* 64, 4, pp. (1242-5)

Watson, CL., Owen, RJ., Said, B., Lai, S., Lee, JV., Surman-Lee, S., & Nichols, G. (2004). Detection of *Helicobacter pylori* by PCR but not culture in water and biofilm samples from drinking water distribution systems in England. *J Appl Microbiol.* 97, pp. (690-698)

Watterworth, L., Topp, E., Schraft, H., Leung, K.T., 2005. Multiplex PCR-DNA probe assay for the detection of pathogenic *Escherichia coli*. *J. Microbiol. Methods*, 60, pp. (93-105)

Wesley, IV., & Miller, GW. (2010). *Arcobacter*: an opportunistic human food-borne pathogen? In: Scheld, W.M., Grayson, M.L., Hughes, J.M. (Eds.), Emerging infections 9, ASM Press, Washington, DC, pp. (185–221)

Westrell, T., Ciampa, N., Boelaert, F., Helwigh, B., Korsgaard, H., Chríel, M., Ammon, A., & Mäkelä, P. (2009). Zoonotic infections in Europe in 2007: a summary of the EFSA-ECDC annual report. Euro Surveill. 14 (3) pii: 19100.

WHO. (2001). Guidelines for Drinking-Water Quality, 2nd Ed. Microbiological Methods, vol.1. World Health Organization, Geneva.

WHO. (2003). Emerging Issues in Water and Infectious Disease. Geneva, Switzerland: WHO Press, World Health Organization.

WHO. (2004). Evaluation of the costs and benefits of water and sanitation improvements at the global level. WHO/SDE/WSH/04.04, Geneva.

Wilkes, G., Edge, T., Gannon, V., Jokinen, C., Lyautey, E., & Medeiros, D., et al. (2009). Seasonal relationships among indicator bacteria, pathogenic bacteria, *Cryptosporidium* oocysts, *Giardia* cysts, and hydrological indices for surface waters within an agricultural landscape. *Water Res*, 43, pp. (2209–2223)

Wilton, S., & Cousins, D. (1992). Detection and identification of multiple mycobacterial pathogens by DNA amplification in a single tube. *PCR Methods Applic.* 1, pp. (269-273)

Wose Kinge, C., & Mbewe, M. (2010). Characterization of *Shigella* species isolated from river Catchments in the North West Province of South Africa. *S. Afr. J. Sci.* 106, 11/12, pp. (1-4)

Wose Kinge, CN., & Mbewe, M. (2011). PCR and sequencing assays targeting *mdh* and *gapA* genes for *Escherichia coli* and *Klebsiella* bacteria species identification in river water from the North West Province of South Africa. *Life Science J.* 8, S1, pp. (104-112)

Yang, F., Yang, J., Zhang, X., Chen, L., Jiang, Y., Yan, Y., Tang, X., Wang, J., Xiong, Z., Dong, J., Xue, Y., Zhu, Y., Xu, X.,  Sun, L., Chen, S., Nie, H., Peng, J., Xu, J., Wang, Y., Yuan, Z., Wen, Y., Yao, Z., Shen, Y., Qiang, B., Hou, Y., Yu, J.,  Jin, Q. (2005). Genome dynamics and diversity of *Shigella* species, the etiologic agents of bacillary dysentery. *Nucleic Acids Res.* 33, 19, pp. (6445–6458)

Younes, M., & Bartram, J. (2001). Waterborne health risks and the WHO perspectives. *Int. J. Hyg. Environ. Health*, 204, pp. (255-263)

# 11

# Identification of Genetic Markers Using Polymerase Chain Reaction (PCR) in Graves' Hyperthyroidism

P. Veeramuthumari and W. Isabel

*PG & Research Department of Zoology and Biotechnology,*
*Lady Doak College, Madurai, Tamil Nadu*
*India*

## 1. Introduction

Thyroid is a butterfly shaped gland composed of two encapsulated lobes, located on either side of the trachea just below the cricoid cartilage. This is connected by thin isthmus and is composed of spherical thyroid follicles, which contain the hormone in colloidal form. $T_3$ and $T_4$ are active hormones secreted under the control of TSH from adenohypophysis of pituitary gland. $T_3$ is three to four fold more potent than $T_4$. It is involved in normal growth and development in children temperature regulation, metabolism, energy production and intelligence in both children adults. It ensures normal growth and development of nervous system [1].

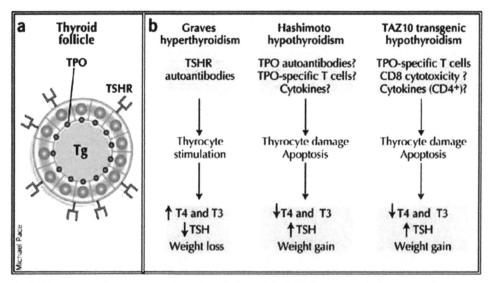

Fig. 1. Diagrammatic representation of variation of thyroid hormones in hypo and hyper thyroidism

The normal range of $T_4$ is suggested to be 77-155nmol/L, T3 to 1.2 -2.8nmol/L ) and TSH to be 0.3-4 mU/L [2]. If the hormone levels are above or below the normal range, it leads to hyperthyroidism or hypothyroidism. The most common hypothyroid condition is Hashimoto's thyroiditis in adults and congenital hypothyroidism in children. Hyperthyroid conditions include Graves' disease, postpartum thyroiditis and thyrotoxicosis factitia.

Hyperthyroidism also leads to a number of complications like heart problems, brittle bones (Osteoporosis), eye problems (Graves' opthalmopathy) (Figure:2).

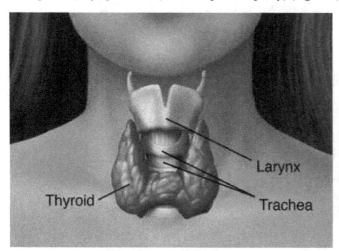
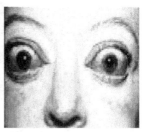

Fig. 2. Symptoms of Graves' disease

Hypothyroidism describes an under active thyroid gland that is producing low level of thyroid hormone. Hypothyroid patients experience a variety of symptoms, including weight gain, intolerance to cold, goiter (enlarged thyroid), dry coarse, skin, fatigue, constipation, decreased heart rate, poor memory and depression.

The most common form of hyperthyroidism is Graves' disease (GD), an autoimmune disorder accounting for 60-80 % of all cases, in which the antibodies produced by immune system stimulates thyroid gland to produce excess of thyroxine. Normally, the immune system uses antibodies to protect against viruses, bacteria and other foreign substances that enter the body system. In GD, the antibodies mistakenly attack the thyroid gland and occasionally the tissues behind the eyes and the skin of lower legs over the shins. Though the exact cause of GD is not known, several factors including a genetic predisposition are likely to be involved **(Figure:3)**.

GD is an organ specific heterogeneous autoimmune disorder associated with T-lymphocyte abnormality affecting the thyroid eyes and skin. GD is also multifactorial disease that develops as a result of complex interaction between genetic susceptibility genes and environmental factors. Human leucocyte antigen (HLA) and cytotoxic T-lymphocyte associated molecule-4 (CTLA-4) are susceptibility candidates. CTLA_4 gene plays an important role in the development of GD, which is located on chromosome 2 q33.

STIMULATING AUTO-ANTIBODIES (Graves' disease)

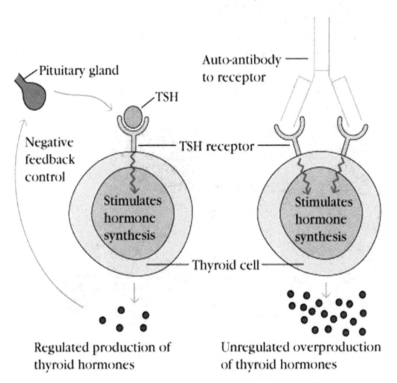

Fig. 3.

## 2. Cytogenetic location of CTLA-4 gene

Cytogenetic Location: 2q33

Molecular Location on chromosome 2: base pairs 204,732,510 to 204,738,682

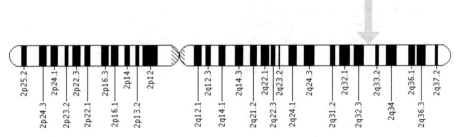

Fig. 4. The CTLA4 gene is located on the long (q) arm of chromosome 2 at position 33.

More precisely, the *CTLA4* gene is located from base pair 204,732,510 to base pair 204,738,682 on chromosome 2.

Activation of T cells requires 2 signals transduced by the antigen specific TCR and co stimulatory ligand such as CD28. CTLA-4, which is expressed on activated T cells, bind to B7 present on antigen presenting cells and functions as a negative regulator of T cell activation. CTLA-4 gene polymorphism confers susceptibility to several autoimmune diseases, such as Graves' disease (GD), Hashimoto's thyroiditis (HT), Addison's disease (AD), Insulin-dependent diabetes mellitus (IDDM), Rheumatoid arthritis (RA) and Multiple sclerosis.

The activity of T cells requires a co stimulatory signal mediated by CD28/B7 interaction. The CTLA-4 gene product delivers a negative signal to T cells and mediates apoptosis. This CTLA-4 gene product is a T cell surface molecule that binds to the B 7 molecule on the antigen presenting cells (APCs). The CTLA-4 gene expression on T cells may affect the course of ongoing immune process. TSH receptor antibody (TRAb) causes Graves' hyperthyroidism.

The GD will go into remission during antithyroid drug (ATD) treatment. Remission of GD is predicted by a smooth decrease in TRAb during (ATD) treatment. Treatment of GD may involve surgery or use of radioactive iodine or use of ATD like propylthiouracil, methimazole and carbimazole. The genetic susceptibility to GD is also conferred by genes in human leucocyte antigen (HLA) and several other genes that are not linked to HLA. The present paper describes the association of GD with the CTLA-4 gene.

The prevalence of hyperthyroidism has been reported to be 3.63% and hypothyroidism to be 2.97% especially the females being more affected by hyperthyroidism [3]. Hence the current study deals with A/G single nucleotide polymorphism (SNP) at position 49 (exon1, codon 17) of the CTLA-4 gene where in Thr/Ala substitution and can be a function related marker. It has been shown to be associated with GD in Caucasians, Japanese, Koreans, Tunisians, Hong Kong Chinese children [2,4,5,6,7,8,9,10] and South Indains [11,12].

The polymorphism cited (A/G polymorphism in exon 1, C/T polymorphism in the promoter, and micro satellite repeat in 3'-untranslated region of exon 4) in CTLA-4 gene have been reported to be associated with autoimmune endocrine disorder.

**A/G polymorphism** at position 49 in exon 1 of the CTLA-4 gene among South Indian population with Graves' hyperthyroidism has revealed the frequencies of the GG genotype and "G" allele to the significantly higher in GD patients. The study has also demonstrated that GD patients had higher frequencies of "G" allele (GG genotype) and lower frequencies of "A" allele (AA genotype) than control group.

Kinjo *et al.*, (2000) have also reported the relationship between the CTLA-4 gene type and severity of the thyroid dysfunction. At diagnosis, free T4 concentrations were shown to be more in patients with the GG genotype and low in patients with the AA genotype. GD patients were reported to have more "G" allele than the control, suggesting that the CTLA-4 GG genotype might induce down regulation of T-cell activation. If the function of CTLA-4 with "G" alleles at position 49 in exon 1 is impaired CTLA-4 function may have d\difficulty in achieving remission.

### Identification of SNP

We can analyze and identify all types of gene SNPs by Polymerase Chain Reaction (PCR) thermal cycler.

## Polymerase Chain Reaction

The polymerase chain reaction (PCR) is a laboratory (in vitro) technique for generating large quantities of a specified DNA. Obviously, PCR is a cell-free amplification technique for synthesizing multiple identical copies (billions) of any DNA of interest, which was developed in 1994 by Karry Mullis (Nobel Prize, 1993). PCR is now considered as a basic tool for any molecular biologist.

## 3. Primer designing [13]

As oligonucleotide primers are useful for polymerase chain reaction (PCR), oligo hybridization and DNA sequencing, proper primer designing is actually one of the most important factors/steps. Various bioinformatics programs are available for selection of primer pairs from a template sequence.

### 3.1 Guidelines for primer design

When choosing two PCR amplification primers, the following guidelines should be considered:

**Primer length:** It is accepted that optimal length of PCR primers is 18-22 bp (Wu et al., 1991)

**Melting temperature (Tm):** It can be calculated using the formula of Wallace et al., 1997, Tm (ºC) = 2(A+T)+4(G+C). The optimal melting temperature for primers ranges between 52-58ºC. Primers with melting temperature above 65ºC should also be avoided because of potential for secondary annealing.

**Primer annealing temperature:** The two primers of a primer pair should have closely matched melting temperatures for maximizing PCR product yield. The difference of 5ºC or more each can lead to no amplification.

$$T_a = 0.3 \times T_m (\text{primer}) + 0.7 \qquad (T_m \text{primer} = 14.9)$$

Where, Tm (primer) = Melting Temperature of the primers, $T_m$ (product) = Melting temperature of the product.

**GC Content:** Primers should have GC content between 45 and 60 percent. GC content, melting temperature and annealing temperature are strictly dependent on one another

.Dimers and false priming because misleading results: Presence of the secondary structures such as hairpins, self dimer produced by intermolecular or intramolecular interactions in primers can lead to poor or no yield of the product.

**Avoid Cross homology:** To improve specificity of the primers it is necessary to avoid regions of homology

### 3.2 Software for primer design

**NETPRIMER** is software used to design and analyze the parameters of designed primer sequences using the following link http://premierbiosoft.com/netprimer/index.html

## 3.3 PCR standardization [13,14]

PCR is a revolutionary technique used in almost all molecular biology experiments. In PCR, the repeated three-step process of denaturation, primer annealing and DNA polymerase extension results in exponential amplification of target DNA. Initially PCR was reported with E.Coli DNA polymerase Klenow fragment in 1985. In 1988, the first report on PCR using thermostable Taq DNA polymerase was published. Since then PCR has been extensively modified and used for various applications such as cloning, sequencing, site-directed mutagenesis, diagnostics, genotyping, genome walking, amplification of RNA after reverse transcription for gene expression analysis amplification of a whole genome, etc.

The central components of a PCR reaction are oligonucleotide primers, thermostable DNA polymerase, target DNA, dNTPs and reaction buffer including $MgCl_2$. When a new PCR has to be developed, suitable primer pairs should be designed based on the target sequence,. Subsequently, the concentration of PCR components and the cycling conditions should be optimized.

**Thermostable enzymes:**

Thermostable enzymes should be selected based on the applications. High fidelity Taq DNA polymerase and proofreading recombinant enzymes are required for the amplification of more than 3 kb target sequence. For a standard PCR, 2 to 5 units of Taq DNA polymerase are recommended for a typical 100µl PCR.

**Deoxynucleoside triphosphate (dNTPs):**

For a standard PCR, 100 to 200 µM concentrations of dNTPs is used. The balanced solutions of all four nucleotides should be used to minimize the error frequency. The concentrations may be increased for Multiplex PCR and Repetitive PCR, where more than one PCR amplicons are expected.

**Template DNA:**

The purity and concentration of the template DNA are critical for a successful PCR amplification. For initial experiments, 0.1 to 200ng of the template DNA, based on the type can be used. For example, if it is a plasmid 0.1 to 1ng is sufficient. If the template is human genomic DNA, upto 200ng can be used.

**Primer concentrations:**

The primer concentration can affect the PCR. If the primer concentration is too low, amplifications will be failed; and if the concentration is too high, non-specific amplification will occur. Therfore, the primer concentration should be optimized empirically between 0.1 to 1µM final concentrations. The most straightforward way of optimizing a PCR with a given primer pair is to change the concentration of $MgCl_2$ or the annealing temperature or both.

**Optimization of primer annealing temperature:**

Optimization of the primer annealing temperature is the most critical step in PCR. The primer designing programs will suggest the Tm of the primers. In general, the annealing temperature should be set 2 to 5ºC below the Tm of the primers. However, some oligonucleotides may not work optically at this temperature and hence the annealing temperature should be optimized using gradient PCR approach.

**Optimization of MgCl₂ concentration:**

Magnesium chloride is an essential component for PCR. It is a cofactor for Taq DNA polymerase. $Mg^{++}$ promotes DNA/DNA interactions and forms complexes with dNTPs that are the actual substrates for Taq polymerase. When Mg++ is too low, primers fail to anneal to the target DNA. When $Mg^{++}$ is too high, the base pairing becomes too strong and the amplicon fails to denature completely when you heat 94°C. MgCl₂ concentration should be optimized for every PCR reaction. All the components of the reaction mixture can bind to magnesium ion, including primers, template, PCR products and dNTPs. Therefore, the concentration of MgCl₂ has to be optimized for a new PCR. The most commonly used concentration of MgCl₂ is 1.5mM and it can be optimized empirically between 1.5 and 4.0mM.

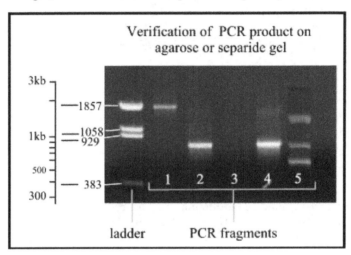

Fig. 5.

### 3.4 Genetic marker [15,16,17,18]

A **genetic marker** is a gene or DNA sequence with a known location on a chromosome that can be used to identify cells, individuals or species. It can be described and observed as a variation which may arise due to mutation or alteration in the genomic loci. A genetic marker may be a short DNA sequence, such as a sequence surrounding a single base-pair change (single nucleotide polymorphism, SNP), or a long one, like minisatellites. For many years, gene mapping was limited in most organisms by traditional genetic markers which include genes that encode easily observable characteristics such as blood types or seed shapes.

### 3.5 Some commonly used types of genetic markers

RFLP (or Restriction fragment length polymorphism)
SSLP (or Simple sequence length polymorphism)
AFLP (or Amplified fragment length polymorphism)
RAPD (or Random amplification of polymorphic DNA)
VNTR (or Variable number tandem repeat)

Microsatellite polymorphism, SSR (or Simple sequence repeat)
SNP (or Single nucleotide polymorphism)
STR (or Short tandem repeat)
SFP (or Single feature polymorphism)
DArT (or Diversity Arrays Technology)
RAD markers (or Restriction site associated DNA markers)
They can be further categorized as dominant or co-dominant.

**Dominant markers** allow for analyzing many loci at one time, e.g. RAPD. A primer amplifying a dominant marker could amplify at many loci in one sample of DNA with one PCR reaction. The dominant markers, as RAPDs and high-efficiency markers (like AFLPs and SMPLs), allow the analysis of many loci per experiment within requiring previous information about their sequence.

**Co-dominant markers** analyze one locus at a time. A primer amplifying a co-dominant marker would yield one targeted product. so they are more informative because the allelic variations of that locus can be distinguished. As a consequence, we can identify linkage groups between different genetic maps but, for their development it is necessary to know the sequence (which is still expensive and is considered one of their down sides). **Eg.** RFLPs, microsatellites, etc.,

### 3.6 Uses of genetic markers

* Genetic markers can be used to study the relationship between an inherited disease and its genetic cause (for example, a particular mutation of a gene that results in a defective protein). It is known that pieces of DNA that lie near each other on a chromosome tend to be inherited together. This property enables the use of a marker, which can then be used to determine the precise inheritance pattern of the gene that has not yet been exactly localized.
* Genetic markers have to be easily identifiable, associated with a specific locus and highly polymorphic, because homozygotes do not provide any information.
* Detection of the marker can be direct by RNA sequencing, or indirect using allozymes.
* Genetic Markers have also been used to measure the genomic response to selection in livestock.
* Natural and artificial selection leads to a change in the genetic makeup of the cell. The presence of different alleles due to a distorted segregation at the genetic markers is indicative of the difference between selected and non-selected livestock.

Hence, SNP (Single nucleotide polymorphism) in Graves' hyperthyroidism is used as marker to identify which mutation is responsible for causing GD and other hereditary diseases.

## 4. Analysis of CTLA-4 A/G polymorphism among South Indian population

### 4.1 Protocol used for A/G single nucleotide polymorphism (SNP) study in Graves' disease

Genomic DNA was prepared from peripheral white cells using standardized protocol. We have analysed CTLA -4 genotypes and allele with PCR. PCR was performed with oligonucleotide primers (Forward, 5' – GCTCTACTTCCTGAAGACCT – 3' and Revers, 5' – AGTCTCACTCACCTTTGCAG – 5')[2]. PCR was performed by initial denaturation 30 sec for 5 min. annealing for 45 sec at 57°C, extension for 30 sec at 72°C, denaturation 30 sec at 94°C (for 20 cycles) and final extension for 7 min at 72°C. The PCR product was confirmed by agarose (1.8%) gel electrophoresis. The presence of G alleles was determined in each subject by PCR amplification of CTLA-4, followed by diffusion with *Bbv1*, which acts on the G variation, but not on the A variation. It a G allele was at position 49, 88/74 bp fragments were obtained. This was confirmed by 2% agarose gel.

## 4.2 Restriction digestion

The amplified CTLA-4 gene should be digested with the restriction enzyme *Bbv1*,which is commercially available. A typical 30µl reaction mix was used . Modify the required volume proportionately.

| | |
|---|---|
| PCR amplified product | – 20. 0µl |
| 10x buffer | - 3.0µl |
| Bbv1(10units/ul) | - 1.0µl |
| Deionized water | - 6.0µl |
| Total | - 30.0µl |

Incubated the reaction mixture at 37°C for 4 hrs and inactivated by heating at 70°C for 10 min. The product was confirmed using 2% agarose gel electrophoresis.

## 4.3 Results

The presence of genomic DNA confirmed by subjecting the agarose gel electrophoresis (0.7%) **(Figure 6)**. The genomic DNA was then subjected to PCR and 162 bp fragments were obtained **(Figure 7)**. The amplified PCR product digested with enzyme ***Bbv1,*** the restriction enzyme acts on the G variation, but not on the A variation. If a G allele was at position 49, 88bp and 74bp fragments were obtained and the fragments were detected by 2% agarose gel electrophoresis **(Figure 8)**.

In the present study, the G/G genotype was observed in 32 (40 %) GD patients and in 26 (32.50 %) individuals of the control group, A/G genotype was found in 37 (46.25 %) patients and in 25 (31.25 %) persons of the control group, A/A genotype was observed in 11 (13.75 %) patients and in 29 (36.25 %) persons of the control group and G allele was found in 50 (62.5%) GD patients and in 38 (47.5 %) persons of the control group, and A allele was found in 30 (37.5 %) GD patients and 42 persons (52.5%) of the  control group **(Table 1)**. There was significant difference (p <0.05) in genotype and allelic frequency between the control group and GD patients. The present study also demonstrates an association between the CTLA-4 gene polymorphism in Graves' disease and with the remission rate of Graves' hyperthyroidism. Among the GD cases studied, only 2% had remission. The frequencies of GG genotype (40 %) and G allele (62.5%) were higher when compared to A/A genotype (13.75%) and A allele (37.5 %) (Table 1).

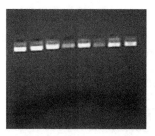

Fig. 6. Confirmation of human genomic DNA

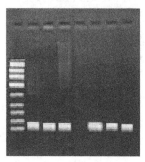

Fig. 7. CTLA-4 gene amplification

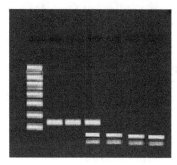

Fig. 8. Restriction analysis of CTLA-4 gene

| GENOTYPE | GD patient (n=80) | Control group (n=80) |
|----------|-------------------|----------------------|
| G/G | 32 (40%) | 26 (32.50%) |
| A/G | 37 (46.25%) | 25 (31.25%) |
| A/A | 11 (13.75%) | 29 (36.25%) |
| Allele | | |
| G | 50 (62.5%) | 38 (47.5%) |
| A | 30 (37.5%) | 42 (52.5%) |

Table 2. Prevalence of CTLA-4 gene genotype and allele frequency among South Indian

## 4.4 Discussion

In the present study genomic DNA was isolated from patients and control groups and was subjected to Agarose gel electrophoresis (0.7%). This enables easy visualization of DNA band patterns. After confirming the presence of genomic DNA, it was subjected to PCR and 162 bp fragments were obtained. The amplified PCR product was digested with enzyme *Bbv*1. The restriction enzyme acts on the G variation, but not on the A variation. If a G allele was at position 49, 88bp and 74bp two fragments were obtained. The PCR products were detected by 2% Agarose gel electrophoresis.

A/G polymorphism at position 49 in exon 1 of the CTLA-4 gene among Madurai population with Graves' hyperthyroidism revealed that the frequencies of the GG genotype and G allele were significantly higher in GD patients. This study has also revealed lower frequency (or absence) of A allele (AA genotype) than the control. CTLA-4 gene polymorphism has been reported to be associated with GD. CTLA-4 molecule is a member of the family of cell surface molecule CD28, which binds to B7. The CTLA-4/B7 complex competes with the CD28/B7 complex and delivers negative signals to the T-cells, which affects T-cell expansion, cytokine production, and immune responses as evidenced by Park *et.al.*[6] in Korean population, Yanagawa *et al.* [9] in Japanese population and Yanagawa *et al.* [8] in Caucasian population. However, we do not know how CTLA-4 gene polymorphisms may contribute to the development of Graves' hyperthyroidism.

Three polymorphism sites (A/G polymorphism in exon 1; C/T polymorphism in the promoter, and micro satellite repeat in the 3'-untranslated region of exon 4) in the CTLA-4 gene have been reported to be associated with autoimmune endocrine disorders. Kinjo *et. al.*, [2] have reported the relationship between the CTLA-4 gene type and severity of the thyroid dysfunction. At diagnosis, free $T_4$ concentrations were shown to be highest in patients with the GG genotype and lowest in patients with the AA genotype. GD patients have more G allele than control, suggesting that the CTLA-4 GG genotype might induce down regulation of T-cell activation. If the function of CTLA-4 with the G alleles at position 49 in exon 1 was impaired CTLA-4 function might have difficulty in achieving remission.

|          | Graves' Disease<br>% (n = 144) | Controls<br>% (n = 110) |
|----------|--------------------------------|-------------------------|
| Genotype |                                |                         |
| G/G      | 50 (34.7)                      | 26 (23.6)               |
| A/G      | 62 (43.1)                      | 46 (41.8)               |
| A/A      | 32 (22.2)                      | 38 (43.6)               |

Table 3. Frequency of the genotype and allele of A/G polymorphism at position 49 in exon 1 of CTLA-4 gene in GD patients and controls among Japanese –population. [2]

Bednarczuk *et. al.*, [4] analysed the association of CTLA-4 A49G polymorphism with Graves' disease in Caucasian and Japanese population. Their study also reveals that, CTLA –4 G allele and G/G genotype confer genetic susceptibility to GD in Caucasian and Japanese population.

The study of Kouki *et. al.*, [5] among patients with GD revealed there were more individuals with G/G (17.8 %GD vs 11.6% of controls) or A/G CTLA-4 exon 1 genotypes (64.4 % GD vs 53.5% control) and significantly fewer individuals with the A/A alleles (17.8 %GD vs 43.9

%control) when compared with controls. According to their findings, the frequency of the G allele was higher in GD patients (50%) than in controls (38.4%) in their population.

There was significant difference between the control group and GD patients both in genotype and allelic frequency. Therefore, in accordance with previously published results, the present study also demonstrates an association between the CTLA-4 gene polymorphism in Graves' disease and with the remission rate of Graves' hyperthyroidism. Among the GD cases studied, only 2% had remission and the frequencies of GG genotype and G allele were higher when compared to A/A genotype and A allele. GD patient with G allele in exon 1 of the CTLA-4 gene were required to continue Anti thyroid drug (ATD) treatment [19] for longer periods to achieve remission. Further studies will be required to determine a clear association of the CTLA-4 gene polymorphism with the remission of GD.

We have studied another gene polymorphism called PKD1 (C/T) at position 4058 in exon 45 which is responsible for causing autosomal polycystic kidney disease (ADPKD) among South Indian.

## 5. Short summary of C/T polymorphism in PKD1 gene

Polycystic kidney disease (PKD) is a group of monogenic disorders that result in renal cyst development in kidney leads to kidney failure. Autosomal dominant polycystic kidney disease (ADPKD) and autosomal recessive polycystic kidney disease (ARPKD) are two forms of PKD, which are largely limited to the kidney and liver, which extends from neonates to old age. ADPKD is a commonly inherited disorder in humans, with a frequency among the general population of 1 in 500. ADPKD caused by mutations in PKD1 gene (85%) located on human chromosome 16p13.3; the remaining 15% are caused by mutations in the PKD2 gene, located on human chromosome 4q21-23. A total of 60 ADPKD patients among South Indian (Madurai) population were analyzed. In genetic study, the genomic DNA was isolated, which would be subjects into PCR (Figure:9) and RFLP analysis (Figure:10). C/T polymorphism at position 4058 in exon 45 of the PKD1 gene among South Indian (Madurai) population with ADPKD revealed that the "TT" "CT" genotype and the frequency of "T" allele was found be significantly (at p=0.001) higher in the patients compared to control subjects. The study was demonstrated that ADPKD patients had higher frequencies of "T" allele and lower frequency of "C" allele than control subjects. The present study also has been supported by Constantinides et al.,[20]. Therefore, the study reveals that there was an association of C/T polymorphism in ADPKD and the prevalence of ADPKD among South Indian (Madurai) population.

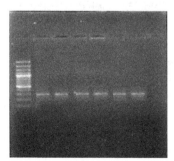

Fig. 9. PKD1 gene amplification

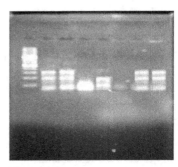

Fig. 10. Restriction digestion PKD1 gene

Hope this chapter will provide an insight on genetic screening of different disease and genetic disorders.

## 6. Acknowledgement

We would like to acknowledge Gunasekaran P, Dr.Sujatha K, Dr.Mahalakshmi A, UGC-NRCBS, School of Biological Science, Madurai Kamaraj University, Madurai, for valuable guidance in standardization of PCR and Primer designing. And also we like to acknowledge Dr.R. Shenbagarathai & faculty in PG & Research Department of Zoology and Biotechnology, Lady Doak College, Madurai, for their continuous support.

## 7. References

[1] Guyton. (1991) Text book of medical physiology, 1091 – 95.
[2] Kinjo Y., Takasu N., Komiya I., Tomoyose T., Takara M., Kouki T., Shimajiri Y., Yabiku K., and Yoshimura H. (2002) Remission of Graves' hyperthyroidism and A/G polymorphism at position 49 in exon 1 of Cytotoxic T- lymphocyte-associated molecule-4 gene. J. of clin. Endocrinol. and Metab. 87(6): 2593-2596.
[3] Velumani A., Kadival GV., Nirmala R. and Lele RD. (2005) Hyperthyroidism. Health screen. 17-21.
[4] Bednarczuk T., Hiromatsu Y., Fukutani T., Jazdzewski K., Miskiewicz P., Osikowska M. and Nauman J. (2003) Association of cytotoxic T-lymphocyte-associated antigen-4 (CTLA-4) gene polymorphism and non-genetic factors with Graves' ophthalmophathy in European and Japanese populations. European J. of Endocrin. 148:13-18.
[5] Kouki T., Sawai Y., Gardine C.A., Fisfalen M-E., Alegre M-L. and Degroot L.J. (2000) CTLA-4 gene polymorphism at Position 49 in Exon 1 reduces the inhibitory function of CTLA-4 and Contributes to the Pathogens of Graves, disease. The J. of Immunol. 165:6606-6611.
[6] Park Y.J., Chung H.K., Park D.J., Kim W.B., Kim S.W., Koh J.J. and Cho B.Y. (2000) Polymorphism in the promotor and exon 1 of the cytotoxic T lymphocyte antigen-4 gene associated with autoimmune thyroid disease in Koreans. Thyroid. 10:453 –459.
[7] Vaidya B., Imrie H. and Perros P. (1999) The cytotoxic T-lymphocyte antigen –4 is a major Graves disease locus. Hum. Mol. Genet. 8:1195-99.

[8] Yanagawa T., Hidaka Y., Guimaraes V., Soliman M. and DeGroot L.J. (1995) CTLA-4 gene polymorphism associated with Graves' disease in a Caucasian population. J. Clin. Endocrinol. Metab. 80:41-45.

[9] Yanagawa T., Taniyama M., Enomoto S., Gomi K., Maruyama H., Ban Y. and Saruta T. (1997) CTLA-4 gene polymorphism confers susceptibility to Graves' disease in Japanese. Thyroid. 7:843-846.

[10] Wang P-W., Liu R-T., Jou S-H.H., Wang S-T., Hu Y-H., Hsieh C-J., Chen M-C., Chen I-Y. and Wu C-L. (2004) Cytotoxic T lymphocyte associated molecule – 4 ploymophism and relapse of Graves' hypethyroidism after Antityroid withdrawal. J. Clin. Endocrinol. Metab. 89(1): 169-173.

[11] Veeramuthumari P, Isabel W, Kannan K. (2009) "A study on the level of serum T.Chol., TGL, HDL, LDL in patients with Graves' hyperthyroid in Madurai population" in Indian J of Endocrinol and Met.

[12] Veeramuthumari P, Isabel W, Kannan K. (2010) "A Study on the level of T3, T4,TSH and the association of A/G Polymorphism with CTLA-4 gene in Grave's Hyperthyroidism among South Indian Population" Indian J of Clin Biochem: 26, 66-69.

[13] Gunasekaran P, Sujatha K. (2010) UGC NRCBS Winter School on Gene Cloning and Expression in Bacteria, Lab Manual. 9-19.

[14] Satyanarayana U, Chakrapani U. ((2006) Biochemistry, Arunabha Sen Book and Allied (P) Ltd. 594-596.

[15] de Vicente, C., T. Fulton (2003). *Molecular Marker Learning Modules – Vol. 1.*. IPGRI, Rome, Italy and Institute for Genetic Diversity, Ithaca, New York, USA.

[16] de Vicente, C., T. Fulton (2004). *Molecular Marker Learning Modules – Vol. 2.*. IPGRI, Rome, Italy and Institute for Genetic Diversity, Ithaca, New York, USA..

[17] de Vicente, C., J-C. Glaszmann, editors (2006). *Molecular Markers for Allele Mining*. AMS (Bioversity's Regional Office for the Americas), CIRAD, GCP, IPGRI, M.S. Swaminathan Research Foundation.

[18] Spooner, S., R van Treuren and M.C. de Vicente (2005). *Molecular markers for genebank management*. CGN, IPGRI, USDA.

[19] Rodriguez S., Quinn F.B., Matthew W. and Ryan W. (2003) Benign thyroid disease. J. Pathol. 33-34.

[20] Constantinides R, Xenophoutos S, Neophyton P, Nomura S, Pierides A and Deltas CC: New amino acid polymorphism, Ala/Val4058, in exon 45 of the polycystic kidney disease 1 gene. Hum Genet 1997, 99:644-647.

# PCR-RFLP and Real-Time PCR Techniques in Molecular Cancer Investigations

Uzay Gormus[1], Nur Selvi[2] and Ilhan Yaylim-Eraltan[3]
*[1]Istanbul Bilim University, Department of Medical Biochemistry,*
*[2]Ege University, Department of Medical Biology*
*[3]Istanbul University, Department of Molecular Medicine*
*Turkey*

## 1. Introduction

### 1.1 Polymerase Chain Reaction (PCR)

The polymerase chain reaction (PCR) is a rapid scientific method for generating a $10^6$-$10^7$-fold increase in the number of copies of discrete DNA or RNA sequences (Boehm,1989; Imboden et al,1993). The use of PCR technology has greatly increased the ability to study on genetic material. PCR is a rapid and reliable molecular biology technique that allows quick replication of mainly DNA, the starting material can be a single molecule of rRNA or mRNA. It was developed by Kary Mullis in 1983, and he was awarded the Nobel Prize in 1993. PCR method is useful in situations of limited amount of DNA sample as in forensics, prenatal testing, because it amplifies a single or a few copies of DNA creating millions of copies of the region(1). The ability to quickly produce large quantities of genetic material has enabled significant scientific advances including DNA fingerprinting and sequencing of the human genome. As PCR technology allows taking specimen of genetic material even from just one cell, copy its genetic material several times, this facilitates genetic studies. Currently, besides research purposes, PCR technology is heavily used in diagnosis and patient management especially for viral diseases such as AIDS and hepatitis. Other than detection of infectious organisms, this technology is also useful for determination of genetic polymorphisms or mutations of individuals (Stahlberg,2011).

The method relies on thermal cycles of repeated heating and cooling of the reaction for DNA melting. Double stranded DNA can be disrupted by heat or high pH, giving rise to single stranded DNA. The single stranded DNA serves as a template for synthesis of a complementary strand by replicating enzymes, DNA polymerases. In order to imitate the accelerated form of DNA replication for a gene region, a special form of DNA polymerase is used. This DNA polymerase should be resistant to the thermal denaturation. Most of the PCR applications employ Taq polymerase, an enzyme isolated from the bacterium Thermus aquaticus, but there are some other heat-stable DNA polymerases used by the same purpose. Most polymerases require short regions of double stranded nucleic acid to initiate synthesis. For in vitro PCR reactions, this can be provided by synthetic oligonucleotides of about 21-25 bp that are complementary to the negatice strand of main DNA molecule. Those oligonucleotide sequences are known as 'primer' and chosen due to the DNA region that we want to amplify. In PCR, two synthetic primers that flank the region of interest are used; one primer is complementary to the negative strand of DNA and second primer to the positive strand. The primers must be oriented that DNA synthesis proceeds across the regions defined by the primers. By this way, only a single region of giant DNA molecule can be amplified. As only one amplification is not enough, PCR is a cyclic process to generate $10^6$-$10^7$-fold increase in a gene region; each PCR cycle contains three steps. Those thermal cycling steps are necessary separate two strands in the DNA double helix at a high temperature by a process called DNA melting. There are three main sequentially repeating steps of PCR:

- *Denaturation* of DNA duplex (94-98°C),
- *Annealing* of primers (37-60°C),
- *Extention* (elongation) of primers by polymerase reaction (~72°C)

In the *denaturation* step, the purpose is to separate strand to be ready for replication, denaturation temperature is higher than the other steps. In the *annealing* step, at a lower temperature, each strand is used as templates for DNA synthesis. The selectivity of PCR results from this step by the usage of primers complementary to the targeted DNA region under specific thermal cycling conditions. After this, there is *extention* step continuing by the heat-stable DNA polymerase to amplify the target DNA region (Boehm,1989).

After 20 cycles of amplification, a million copies of DNA can be generated from a single copy. After several rounds of amplification (about 40 times), the PCR product is analysed on an agarose gel an sis abundant enough to be detected with an ethidium bromide stain.

After this stage, to detect the changes on the DNA sequence, the classical PCR-RFLP method (the next heading) can be used. But also specific DNA sequences can be detected without opening the reaction tube (Higuchi, 1992). Recently, after first preliminary studies the technique developed to get both structural and quantitative informations about the amplified DNA region by real-time PCR devices using flourescent dyes, as we will mention in following headings.

## 1.2 PCR-Restriction Fragment Length Polymorphism (PCR-RFLP) method

In contemporary, there are several forms of PCR that are extensively modified to perform a wide array of genetic manipulations. PCR-RFLP (PCR-restriction fragment length polymorphism) is one of those that was preliminary to most of classified PCR methods. RFLP is a technique referring to a difference between restriction enzyme sites on DNA samples, broken into pieces (digested) by those restriction enzymes and the resulting fragments are separated according to their lengths by gel electrophoresis.

Restriction endonucleases are specific enzymes that can cleave specific nucleotide sequences; because of that property, it is possible for them to discriminate nucleotide changes in DNA. Sometimes they can effect the loci other than the target one, but the important part of the procedure is the possible polymorphism or mutation loci to be detected whether the cleavage site is intact or not. If there is a change in the cleavage site of restriction endonuclease, it will not cleave the site, or by addition of the mutation, there may ocur a previously not existing cleavage site.

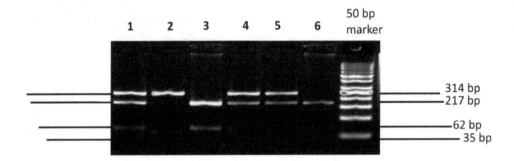

Fig. 1. An example of RFLP results from our laboratories. As shown in the figure, there is a 50 bp marker to compare with our own results and detect the basepair (bp) length. After treatment and incubation with specific restriction enzyme:

- The uncut homozygote cases (having the same alleles) were expected to be having only one 314 bp band (as in number 2).
- The cut homozygote cases (having the same alleles) were expected to be having three bands of 217, 62, 35 bp (as in number 3 and 6)
- The heterozygote cases (having two separate alleles) were expected to be having four bands of 252, 217, 62, 35 bp (as in number 1,4 and 5)

## 1.3 Types of PCR

- **Conventional PCR:** This is the DNA-based PCR, primers target specific sequences on DNA and amplification follows the usual steps of denaturation, annealing and elongation.
- **Reverse transcription-PCR:** mRNA or rRNA can be the main material to be amplified. The first step is the enzymatic 'reverse transcriptase' reaction to transcribe RNA to cDNA. Subsequent steps are similar to conventional PCR steps (Tania et al, 2006).
- **Asymmetric PCR:** It can be used for generation of single strand for sequencing studies. This can be done by adjusting primer concentrations to favor one strand; by this way after first cycles, only the strand complementary to the first strand continues to be copied.
- **Nested PCR:** In this type of PCR, there are two stages of the procedure; in the first part, by using a set of primers, a fragment is amplified. After this, by using another primer set, a sub-region of the previously amplified region is re-amplified. Main aim is to increase sensitivity and specificity.
- **Real-time PCR.**

## 2. Real-time PCR

Real-time PCR (PCR with real time) is also known as kinetic PCR, QPCR, QRT- PCR. Automated thermal cycling devices have been improved by using Taq DNA polymerase which is thermostable and continued to be developed by fluorescence luminescence techniques( Higuchi et al,1992; Logan j et al, 2009.). Real-time PCR is easy to perform, providing reliable results with high accuracy as well as rapid quantification. Quantification of polymorphic DNA regions and genotyping single nucleotide polymorphisms are detected by using the real-time PCR reaction. For gene expression analyses, the mRNA levels can be done quantitatively by reverse transcriptase–PCR (RT-PCR) reaction (Tanie Eet al, 2006). By this way, it is possible to monitor gene outputs numerically in many different fields, from the drug-resistant tumor cells to the chemotherapy scanning and also to the molecular determination of tumor stages. The use of gene expression analysis is getting increased in many notable fields of biological research. Gene profiling opens new possibilities to classify the disease into subtypes and guide a differentiated treatment.

This method has been preferred especially in the samples, the analysis of which cannot be possible, or in the samples, the cytogenetic analysis of which are turned out as auxiliary techniques to the molecular analysis. Therefore, it has became one of the indispensable methods. The introduction of real-time PCR technology has significantly improved and simplified the quantification of nucleic acids, and this technology has become a valuable tool for many scientist working in different disciplines. Especially in the field of molecular diagnosis, real-time PCR-based assays took their advantage (Pfaffl, 2004).

## 2.1 Real-time PCR protocols

Real-time PCR has been preferred as one of the favored methods in molecular studies and in routine analyses, since the process takes short time as 20-30 minutes,it provides fast heating and cooling cycles of 30-40 times, in addition to these, it benefits the control of PCR reaction on a computer monitor (Wittwer, 1997). High sensitivity of real-time PCR makes the technique applicable to very small samples, such as fine needle aspirates. Real-time PCR instruments can simultaneously amplify and detect, eliminating the need to open tubes containing PCR products and therefore reducing the risk of future contamination (Lyon, 2009). Additionally nested PCR and touchdown PCR can be performed using real-time PCR Machine. There are various real-time PCR machines that are used mostly in laboratory experiments:

- ABI Prism 7700
- LightCycler2/Lightcycler 480 Probes (Roche, Mannheim, Germany)
- i-cycler (BioRad)

## 2.2 Probing techniques

Today, fluorescence is exclusively used as the detection method in real-time PCR. The fluorescent reporters can be divided into two categories: nonspecific and sequence-specific labels (Wilhelm, 2003).

**Nonspecific labels**: These are DNA-binding dyes such as SYBR Green I ( Wittwer et al,1997; Zipper et al, 2004.) and BEBO (Bengtsson et al, 2003), which become strongly fluorescent when they are bound to double-stranded DNA. SYBR Green I binds all double-stranded DNA molecules regardless of their sequence. The Double-stranded DNA bindind dye SYBR Green I is proven to be effective. Maximum excitation of SYBR Green I dye occurs at 497 nm. Maximal emission of DNA stained with SYBR Green I occurs at 521 nm. The specifity and sensitivity of SYBR Green I detection can be monitored by performing a melting curve analysis after using the amplification reaction with external standard.

Differentiation of single point mutant alleles from wild type allele is not possible with SYBR Green I but it is possible to detect small deletions/insertions (10 to 20 bp).

### 2.2.1 Hybridization probes (pair of sequence-specific, single-labeled probes)

Sequence-specific probes are based on oligonucleotides or their analogs that one or two fluorescent dyes are coupled.

There are some types of probes with two dyes (Holland et al 1991; Tyagi et al, 1996; Tyagi et al, 1998; Caplin et al 1999): a) hydrolysis probes (TaqMan® probes), b) molecular beacons, c) hybridization probes.

a.  **Hydrolysis probes:** This probe is a single oligonucleotide labeled with two different fluorophores. The fluorophore near the 3' end(acceptor) acts as a fluorescence emission "quencher" of the other one near the 5' end(donor) (Holland et al 1991). As soon as Taq DNA polymerase hydrolyzes the probe via its 5' exonuclease activity during a combined annealing/extension step, the 5' fluorophore (donor) is liberated. Therefore, its emission can no longer be suppressed by the quencher and can be measured in the fluorimeter. TaqMan real-time PCR is one of the two types of quantitative PCR methods, and uses a fluorogenic probe which is a single stranded oligonucleotide of 20-26 nucleotides and is designed to bind only the DNA sequence between two PCR primers. In this case, two primers with a preferred product size of 50-150 bp, a probe with a fluorescent reporter or fluorophore such as 6-carboxyfluorescein (FAM) and tetrachlorofluorescin (TET) and quencher such as tetramethylrhodamine (TAMRA) covalently attached to its 5' and 3' ends are required, respectively (Giller et al, 2011).

b.  **Hybridization probes:** In this case, there are two oligonucleotides that hybridize to adjacent internal sequences of the same amplicon (Witther et al, 2011). For instance, the 5' oligonucleotide (donor) has a fluorescence in label at its 3' end. The 3' oligonucleotide(acceptor) has either LightCycler-Red 640 or LightCycler-Red 705 at its 5' end. Only after hybridization to the template DNA, two probes come in close proximity, resulting in fluorescence resonance energy transfer (FRET) between the two fluorophores. During FRET, fluorescein, the donor fluorophore, is excited especially by the light source of the LightCycler Instrument, and part of the excitation energy is transferred to either LightCycler-Red 640 or LightCycler-Red 705, the acceptor fluorophores. Emitted fluorescence of these acceptor fluorophores are then measured by the LightCycler Instrument. Specific detections are performed with these probes. For example, the mutation detections are analysed via the external and internal standards .

c.  **Molecular Beacon Probes:** A molecular beacon is one oligonucleotide labeled with two different fluorophores, an acceptor and a donor. Due to the specific secondary structure formed by the oligonucleotide (beacon), acceptor (quencher) and donor dyes are in close proximity. A molecular beacon unfolds while binding to the growing PCR product, thereby separating the dyes and enhancing the fluorescence of the donor dye. Four different fluorophores can be designed to detect different point mutations simultaneously (Vincent et al, 2005 ).

### 2.3 Melting curve analysis

At the beginning of a melting curve analysis, the reaction temperature is low and the fluorescence signal is high. As the temperature steadily increases, the fluorescence will suddenly drop as the reaction reaches the melting point ($T_m$) of each DNA fragment. More specific analysis of PCR reactions can be performed with SYBR Green I because of its specific melting behaviour, identification/differentiation of multiple specific PCR products (multiplex PCR ) with SYBR Green I, genotyping and mutation analyses with hybridization probes. Melting Curve Analysis has many advantages (Wittwer et al, 2009). Just like gel

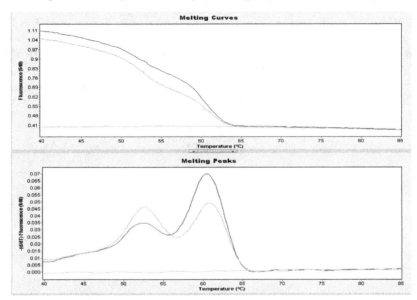

Fig. 2. Heterozygote result indicating two melting curves (53.0 °C and 62.0 °C)

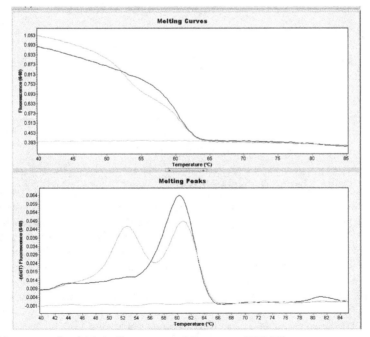

Fig. 3. Wild type result which indicate one melting curve (62.0 °C)

electrophoresis, melting point analysis permits clear identification of the amplicon, since each PCR product possesses a characteristic melting point. Moreover, nonspecific products (primer dimers) can also be identified by this method. If performed with hybridization probes, melting point analysis can also detect point mutations. For instance, the acquired Janus Kinase 2(JAK2) V617F point mutation can be found in more than 90% patient with polycytaemia, and in 50% of patients with other chronic myeloproliferative diseases. For instance in the figures 2-4, our own laboratory results are given. Myeloproliferative neoplasms JAK2V617F-mutation analysis results are shown as melting curve analyses. The genotype is identified by running a melting curve with specific melting points (Tm).

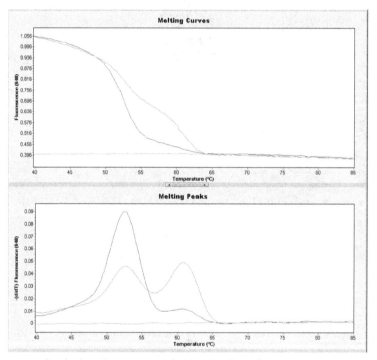

Fig. 4. Mutant result which indicate one melting curve (53.0 °C)

## 2.4 High-Resolution Melting Analysis (HRMA)

High resolution melting is a post-PCR-based method for detecting DNA sequence variation by measuring changes in the melting of a DNA duplex (Martin-Nunez et al, 2011). Melting analysis using new instruments have been designated for high-resolution melting curve analysis (HRM or HRMA) based on its ease of use, simplicity, flexibility, cost-effectivity, nondestructive nature, superb sensitivity, and specificity (Vossen et al, 2009). It enables researchers to rapidly detect and categorize genetic mutations and single nucleotide polymorphisms(SNPs), identify new genetic variants without sequencing (gene scanning) or determine the genetic variation in a population (e.g. viral diversity) prior to sequencing. SYBR®Green I is introduced into a sensitive conventional dye for PCR product melting analysis. High-resolution melting analysis have been used clinically to detect somatic changes in select exons of oncogenes such as *EGFR*,53 *KRAS*,54 *PDGFRA*,55 *KIT*,56 *BRAF*,57 and*TP53* ( Bastien et al, 2008).

## 2.5 Gene expression analysis

Conventional microarrays have limitations in flexibility, speed, cost, and sensitivity. Gene expression analysis by microarray techniques and real-time PCR offers new possibilities to classify malignant tumors, such as lymphomas, into more distinct subtypes for diagnosis and treatment (Schmit et al, 2010; Bagg et al, 1999; Stahlberg et al, 2005. ). The study of biological regulation usually involves gene expression assays and requires quantification of RNA frequently. In the past, conventional gel- or blot-based techniques were used for these assays. However, these techniques often have limitations in speed, sensitivity, dynamic range, and reproducibility required by current experimental systems. In contrast, real-time PCR methods, can easily meet these requirements. Reverse transcription PCR (RT-PCR) is a common and powerful tool for highly sensitive RNA expression profiling. Quantification by real-time PCR may be performed as either absolute measurements using an external standard, or as relative measurements, comparing the expression of a reporter gene with that of a presumed constantly expressed reference gene (Stahlberg et al, 2005).

A flow- chart, represents the steps of Real-time PCR and its applications, is given in Figure 5.

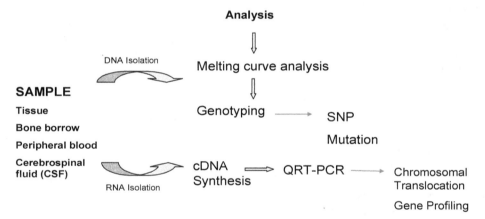

Fig. 5. Real-time PCR Flow-chart

## 2.6 Epigenetic studies with PCR

Epigenetic information is usually lost during the PCR because of the insensitivity of DNA polymerase, it cannot distinguish between methylated and unmethylated cytosines. After PCR, any methylated allele will be extremely diluted. Therefore, something must be done to preserve methylated form of DNA. Treatment with sodium bisulfite will deaminate cytosine to uracil, the rate of deamination of 5-methylcytosine to thymine is slower than the conversion of cytosine to uracil, thus it is assumed that the only cytosines remaining after sodium bisulfite treatment are derived from 5-methylcytosines. By this way, during subsequent PCR cycles, uracil residues are replicated as thymine residues, and 5-methylcytosine residues are replicated as cytosines. The efficiency of the method is about

99% in appropriate conditions, but this method needs intense attention while chosing primers and arranging study conditions(Gulley et al.).

## 3. PCR-based studies in cancer research

The advances in molecular techniques provide new molecular targets for diagnosis and therapy of cancer. These advances can provide both researchers and clinicians with precious information concerning the behavior of tumors. Therefore, these tumors can detect at earlier stages when the tumor burden is smaller and be potentially more curable currently. After the human genome project has completed, the application of highthroughput technologies for polymorphism detection for explaining molecular mechanism for complex disease has created very important opportunities (Khoury,1997).

Single nucleotide polymorphisms (SNPs) offers widespread use in gene mapping of genetic disorders, in the delineation of genetic influences in multifactorial diseases such as cancer, cardiovascular disease, in haplotype mapping, and as genetic markers to predict responses to drugs (Riddick et al, 2005). However, for example, there are some inconsistent results regarding the relationship between the presence of polymorphic forms of genes encoded detoxifying enzymes and chemotherapeutic response. It has been reported that the genetic polymorphism analysis in peripheral blood may not be enough representative for the status in tumour tissue. For instance, Uchida et al reported that individuals heterozygous for the 28-bp polymorphism in thymidylate synthase (TS) gene may have increased risk for cancer that are homozygous for this polymorphism due to loss of one allele during carcinogenesis(Uchida et al, 2004). They also showed that the response to 5-FU-based chemotherapy in these cases was comparable to cases where the individual was homozygous. Therefore, it may be excellent to determine the genotype of polymorphisms in tumour cells than in peripheral blood.

Some data obtained from combined genotype studies have demonstrated that these data may have significance for models of cancer prognosis or treatment. But, many researchers suggest that larger studies will be needed also to investigate the effect of specific treatment modalities in cancer. While investigating the post-initiation stages of cancer, four basic parts can be dedicated to gene polymorphisms affecting: (a)growth control of cell (cell proliferation, differentiation and death); (b)factors involved in tumour invasion and metastasis (immune and inflammatory responses, extracellular matrix remodelling, angiogenesis and cell adhesion); (c)effects of hormones and vitamins on growing tumours; (d)outcome of cancer therapy (cancer pharmacogenetics) (Loktionov, 2004). Quantitation of gene expression in tumor or host cells has another an enormous importance for investigating the gene patterns responsible for cancer development, progression and response or resistance to therapy.

Analysis of transcriptional activity of tumor cells or detection of possible new tumor markers by polymerase chain reaction (PCR), reverse transcriptase PCR (RT-PCR) and quantitative reverse transcriptase polymerase chain reaction (qRT-PCR) techniques have the potential to change cancer diagnosis and treatment (Mocellin, 2003). New molecular tecniques for diagnosis offers the promise of accurately matching patient with treatment. It has been shown that there is a resultant significant effect on improved disease outcome. Currently, the real-time reverse transcription polymerase chain reaction (qRT-PCR), has a potential to become an important analytical tecnique for the mRNA detection in tissue biopsies or body fluids. qRT-PCR is especially promising in prognostic assays and monitoring response to treatment for cancer patients. It is known that histopathological staging in cancer defines patient prognosis. However, there are some limitations in the prognostic heterogeneity of patients within a given tumour stage. According to this view, not all patients with lymph node-negative are treated and not all patients with lymph node-positive tumours die from their cancer. So, more accurate staging protocols are needed for detection clinical tumour staging by using molecular techniques.

Gene expression analysis is one of the most important parameter that utilises the qRT-PCR assay's potential for generating quantitative data (Skrzypski, 2008; Schuster et al, 2004). It is reported that the detection of disseminated tumor cells in peripheral blood obtained from colorectal cancer patients by RT-PCR could be an effective method for identifying patients for adjuvant therapy. It is known that the mRNA for prostate specific antigen (PSA) is expressed only by prostatic cells. RT-PCR are suggested as a potentially more sensitive assay for the detection of cells expressing PSA mRNA in peripheral blood or in extraprostatic tissues. Some studies suggest that the molecular detection of circulating tumor cells (CTC) and micrometastases may help develop new prognostic markers in patients with solid tumors (Ghossein et al, 2000). It has been reported that prostatic tissue specific markers and melanoma related transcripts were detected by RT PCR in the peripheral blood, bone marrow and lymph nodes of patients with localized and advanced tumors. Currently, many reliable methods emerged with fast and efficient mechanisms for screening and monitorizing large populations for genetically linked traits and for cancer-related genes discovery.

In addition to gene expression profiling, real-time PCR is also useful to detect chromosomal aberrations. Non-random chromosomal translocations are frequently associated with a variety of cancers, particularly hematologic malignancies and childhood sarkomas (Peter et al, 2006). For example t(15,17) translocation is found only in the leukemic cells. Only in patients with acute promyelocytic leukemia (APL) and the other forms of leukemia, t(1;19) translocation is found with B-cell precursor acute lympoblastic leukemia (ALL). Quantitative analysis provide small number of remaining malignant cells (minimal residual disease, MRD) in patients to be revealed whose disease is in a clinical remission. Therefore, quantitative results are very important in terms of detection in malignancies and MRD. For example, BCR-ABL quantification monitors MRD and therapy of chronic myelogenous leukemia (Lyon et al, 2009). Using the real-time PCR Instrument as a closed tube, rapid amplification and real-time fluorescence detection system, for example quantitative measurement of the BCR-ABL expression level can be performed with a minimum risk of cross contamination. Relative expression levels of different samples may be calculated by standardizing the amount of BCR-ABL transcripts in a sample to the amount of an endogenous expressed housekeeping gene. The values for BCR-ABL and housekeeping gene for each sample are calculated by the real-time PCR software by the comparing the crossing points to the standard curve. A normalized target value (the ratio of BCR-ABL/housekeeping ) is then derived by dividing the amount of BCR-ABL by the amount of housekeeping gene. The chromosomal aberration examples in various leukemia types can be detected by RNA quantification, shown in Table 1. On the other hand, melting analysis of the PCR product or the probe is used to confirm detection of the correct product.

| ACUTE LYMPHOBLASTIC LEUKEMIA (ALL) | | | ACUTE NON-LYMPHOBLASTIC LEUKEMIA (ANLL) | |
|---|---|---|---|---|
| t(9;22) | BCR-ABL | Translocation | t(15;17) | PML-RARα Translocation |
| t(1;19) | E2A-PRL | Translocation | t(8,21) | AML1-ETO Translocation |
| t(12;21) | TEL-AML1 | Translocation | inv (16) | CBFβ- MYH11 Inversion |
| t(4;11) | MLL-AF4 | Translocation | **CRONIC MYELOID LEUKEMIA (KML)** | |
| Multidrug resistance 1 (MDR1) | | | t(9;22) | BCR-ABL Translocation |

Table 1. The chromosomal aberrations that can be detected by RNA quantification.

## 4. References

Adam Bagg, Bhaskar Kallakury. Molecular pathology of Leukemia and Lymphpma. *Am.J.Clin.Pathol.* 112, 76-92, 1999.

Alexandre Loktionov. Common gene polymorphisms, cancer progression and prognosis *Cancer Letters* Volume 208, Issue 1, 10 May 2004, Pages 1-33

Anders Stahlberg†, Neven Zoric, Pierre Åman and Mikael Kubista Quantitative real-time PCR for cancer detection: the lymphoma case: *Expert Rev. Mol. Diagn.* 5(2), 2005.

Bastien R, Lewis TB, Hawkes JE, Quackenbush JF, Robbins TC, Palazzo J, Perou CM, Bernard PS: High-throughput amplicon scanning of the TP53 gene in breast cancer using high-resolution fluorescent melting curve analyses and automatic mutation calling. *Hum Mutat*, 29:757–764, 2008.

Bengtsson M, Karlsson JH, Westman G, Kubista M. A new minor groove binding asymmetric cyanine reporter dye for realtime PCR. *Nucleic Acids Res.* 31(8), E45, 2003.

Caplin BE, Rasmussen RP, Bernard PS, Wittwer CT. LightCyclerTM hybridization probes – the most direct way to monitor PCR amplification and mutation detection. *Biochemica* 1, 5–8, 1999.

Corlnne D. Boehm. Use of Polymerase Chain Reaction for Diagnosis of Inherited Disorders. *CLIN. CHEM.* 35/9, 1843-1848 (1989).

Devita VT, Hellman S, Rosenberg SA. (2005). Cancer princinles and Practise of Oncology.7th Edition, 2005.

Elaine Lyon and Carl T. Wittwer. LightCycler Technology in Molecular DiagnosticsJournal of Molecular Diagnostics, Vol. 11, No. 2, 2009.

Ghossein RA, Carusone L, Bhattacharya S. Molecular detection of micrometastases and circulating tumor cells in melanoma prostatic and breast carcinomas. *In Vivo.* 2000 Jan-Feb;14(1):237-50.

Gillet JP, Gottesman MM. Advances in the molecular detection of ABC transporters involved in multidrug resistance in cancer. *Review.Curr Pharm Biotechnol.* 12(4):686-92, 2011

Gulley ML, Shea TC, Fedoriw Y. Genetic tests to evaluate prognosis and predict therapeutic response in acute myeloid leukemia. *J Mol Diagn.* 2010 Jan;12(1):3-16. Epub 2009 Dec 3.

Higuchi, R., Dollinger, G., Walsh, P. S., and Griffith, R. "Simultaneous amplification and detection of specific DNA sequences." *Biotechnology* 10:413–417,1992.

Holland PM, Abramson RD, Watson R, Gelfand DH. Detection of specific polymerase chain reaction product by utilizing the 5′-3′ exonuclease activity of Thermus aquaticus DNA polymerase. *Proc. Natl Acad. Sci.* 88(16), 7276–7280, 1991.

Imboden P, Burkart T, Schopfer K. Simultaneous detection of DNA and RNA by differential polymerase chain reaction (DIFF-PCR). *PCR Methods Appl.* 1993 Aug;3(1):23-7.

Khoury, M. J. (1997) Genetic epidemiology and the future of disease prevention and public health. *Epidemiol. Rev.* 19, 175–180.

Logan J, Edwards K, Saunders N. Real-Time PCR: Current Technology and Applications, 2009.

Martín-Nunez GM, Gómez-Zumaquero JM, Soriguer F, Morcillo S. High resolution melting curve analysis of DNA samples isolated by different DNA extraction methods. *Clin Chim.Acta*, 2011.

Michael W. Pfaffl, Quantification strategies in real-time PCR, Chapter 3 pages 87 – 112, 2004.

Mocellin S, Rossi CR, Pilati P, Nitti D & Marincola FM. (2003). Quantitative real-time PCR: a powerful ally in cancer research. *Trends in Molecular Medicine.* 9, 5, 189-195.

Peter D. Aplan, Causes of oncogenic chromosomal translocation,Trends in Genetics, 22(1), 2006.

Riddick DS, Lee C, Ramji S, Chinje EC, et al. (2005). Cancer chemotherapy and drug metabolism. *Drug Metab. Dispos.* 33: 1083-1096.

Schmit A, Molecular markers in hematology and oncology, 22;99(19):1143-52, 2010.

Schuster R, Max N, Mann B, Heufelder K, Thilo F, Gröne J, Rokos F, Buhr HJ, Thiel E & Keilholz U. (2004). Quantitative real-time RT-PCR for detection of disseminated tumor cells in peripheral blood of patients with colorectal cancer using different mRNA markers. *Int J Cancer.*10, 108, 2, 219-27.

Skrzypski M. (2008). Quantitative reverse transcriptase real-time polymerase chain reaction (qRT-PCR) in translational oncology: lung cancer perspective. *Lung Cancer.* 59, 2, 147-54.

Ståhlberg A, Kubista M & Aman P. (2011). Single-cell gene-expression profiling and its potential diagnostic applications. *Expert Rev Mol Diagn.*11, 7, 735-40.

Tania N, Hands RE & Bustin SA. (2006). Quantification of mRNA using real-time RT-PCR. *Nature Protocols.* 3, 1559-1581.

Tyagi S & Kramer FR. (1996). Molecular beacons: probes that fluorescence upon hybridization. *Nature Biotechnol.* 14, 3, 303–308.

Tyagi S, Bratu DP & Kramer FR. (1998). Multicolor molecular beacons for allele discrimination. *Nature Biotechnol.* 16, 1, 49–53.

Uchida K, Hayashi K, Kawakami K, Schneider S, Yochim JM, Kuramochi H, Takasaki K, Danenberg KD & Danenberg PV. (2004). Loss of heterozygosity at the thymidylate synthase (TS) locus on chromosome 18 affects tumor response and survival in individuals heterozygous for a 28-bp polymorphism in the TS gene. *Clin Cancer Res.* 10, 433–439.

Vossen RH, Aten E, Roos A & den Dunnen JT. (2009). High-resolution melting analysis (HRMA): more than just sequence variant screening. *Huma mutant.* 30, 860-6.

Wilhelm J & Pingoud A. (2003). Real-timepolymerase chain reaction. *Chembiochem.* 4, 11, 1120–1128.

Wittwer CT & Farrar JS. PCR Troubleshooting and Optimization, *Hybridization Probes in PCR*, 2011.

Wittwer CT & Kusukawa N. Real-time PCR and melting analysis. (2009). *Molecular Microbiology: Diagnostic Principles and Practice* (ed 2).

Wittwer CT, Herrmann MG, Moss AA & Rasmussen RP. (1997). Continuous fluorescence monitoring of rapid cycle DNA amplification. *Biotechniques.* 22, 130–138.

Zipper H, Brunner H, Bernhagen J & Vitzthum F. (2004). Investigations on DNA intercalation and surface binding by SYBR Green I, its structure determination and methodological implications. *Nucleic Acids Res.* 32, 12, 103.

# Permissions

The contributors of this book come from diverse backgrounds, making this book a truly international effort. This book will bring forth new frontiers with its revolutionizing research information and detailed analysis of the nascent developments around the world.

We would like to thank Patricia Hernandez-Rodriguez, for lending her expertise to make the book truly unique. She has played a crucial role in the development of this book. Without her invaluable contribution this book wouldn't have been possible. She has made vital efforts to compile up to date information on the varied aspects of this subject to make this book a valuable addition to the collection of many professionals and students.

This book was conceptualized with the vision of imparting up-to-date information and advanced data in this field. To ensure the same, a matchless editorial board was set up. Every individual on the board went through rigorous rounds of assessment to prove their worth. After which they invested a large part of their time researching and compiling the most relevant data for our readers. Conferences and sessions were held from time to time between the editorial board and the contributing authors to present the data in the most comprehensible form. The editorial team has worked tirelessly to provide valuable and valid information to help people across the globe.

Every chapter published in this book has been scrutinized by our experts. Their significance has been extensively debated. The topics covered herein carry significant findings which will fuel the growth of the discipline. They may even be implemented as practical applications or may be referred to as a beginning point for another development. Chapters in this book were first published by InTech; hereby published with permission under the Creative Commons Attribution License or equivalent.

The editorial board has been involved in producing this book since its inception. They have spent rigorous hours researching and exploring the diverse topics which have resulted in the successful publishing of this book. They have passed on their knowledge of decades through this book. To expedite this challenging task, the publisher supported the team at every step. A small team of assistant editors was also appointed to further simplify the editing procedure and attain best results for the readers.

Our editorial team has been hand-picked from every corner of the world. Their multi-ethnicity adds dynamic inputs to the discussions which result in innovative outcomes. These outcomes are then further discussed with the researchers and contributors who give their valuable feedback and opinion regarding the same. The feedback is then collaborated with the researches and they are edited in a comprehensive manner to aid the understanding of the subject.

Apart from the editorial board, the designing team has also invested a significant amount of their time in understanding the subject and creating the most relevant covers. They scrutinized every image to scout for the most suitable representation of the subject and create an appropriate cover for the book.

The publishing team has been involved in this book since its early stages. They were actively engaged in every process, be it collecting the data, connecting with the contributors or procuring relevant information. The team has been an ardent support to the editorial, designing and production team. Their endless efforts to recruit the best for this project, has resulted in the accomplishment of this book. They are a veteran in the field of academics and their pool of knowledge is as vast as their experience in printing. Their expertise and guidance has proved useful at every step. Their uncompromising quality standards have made this book an exceptional effort. Their encouragement from time to time has been an inspiration for everyone.

The publisher and the editorial board hope that this book will prove to be a valuable piece of knowledge for researchers, students, practitioners and scholars across the globe.

# List of Contributors

**H. W. Araujo-Torres**
Conservation Medicine Lab., Centro de Biotecnología Genómica del Instituto Politécnico Nacional, Cd. Reynosa, Tamps, Mexico
Centro de Investigación en Ciencia Aplicada y Tecnología Avanzada del Instituto Politécnico Nacional, Altamira, Tamps, Mexico

**J. A. Narváez-Zapata**
Industrial Biotechnology Lab., Centro de Biotecnología Genómica del Instituto Politécnico Nacional, Cd. Reynosa, Tamps, Mexico

**MS. Puga-Hernández**
Laboratorio Estatal de Salud Pública de Tamaulipas, Cd. Victoria, Tamps, Mexico

**J. Flores-Gracia**
Instituto Tecnológico de Ciudad Victoria, Cd. Victoria, Tamps, México

**M. G. Castillo-Álvarez and M. A. Reyes-López**
Conservation Medicine Lab., Centro de Biotecnología Genómica del Instituto Politécnico Nacional, Cd. Reynosa, Tamps, Mexico

**Ayse Gul Mutlu**
Mehmet Akif Ersoy University, Arts and Sciences Faculty, Department of Biology, Burdur, Turkey

**Na Liu, Jianxin Niu and Ying Zhao**
Department of Horticultural, Agricultural College of Shihezi University, Shihezi, People's Republic of China

**Nikola Tanic and Jasna Bankovic**
Institute for Biological Research "Sinisa Stankovic", University of Belgrade, Belgrade

**Nasta Tanic**
Institute of nuclear Sciences "Vinca", Belgrade, Republic of Serbia

**Ursula Theocharidis and Andreas Faissner**
Ruhr-University Bochum, Germany

**Tammam Sipahi**
Department of Biophysics, Medical Faculty, Trakya University, Edirne, Turkey

**Nasim Danaei**
Department of Health and Nutrition, Tabriz University of Medical Sciences, Tabriz, Iran

**Morteza Seifi and Asghar Ghasemi**
Laboratory of Genetics, Legal Medicine Organization of Tabriz, Tabriz, Iran

**Siamak Heidarzadeh**
Division of Microbiology, School of Public Health, Tehran University of Medical Sciences, Tehran, Iran

**Mahmood Khosravi**
Hematology Department of Medicine Faculty, Guilan University of Medical Sciences, Rasht, Iran

**Atefeh Namipashaki**
Department of Biotechnology, School of Allied Medical Sciences, Tehran University of Medical Sciences, Tehran, Iran

**Vahid Mehri Soofiany**
Faculty of Medicine, Shahid Behesti University of Medical Sciences, Tehran, Iran

**Ali Alizadeh Khosroshahi**
Jarrah Pasha Medicine Faculty of Istanbul, Istanbul, Turkey

**P. S. Shwed**
Biotechnology Laboratory, Environmental Health Sciences and Research Bureau, Environmental and Radiation Health Sciences Directorate, Healthy Environments and Consumer Safety Branch, Health Canada, Canada

**Fousseyni S. Touré Ndouo**
Medical Parasitology Unit, Centre International de Recherches Médicales de Franceville (CIRMF), Franceville, Gabon

**C. N. Wose Kinge and N. P. Sithebe**
Department of Biological Sciences, School of Environmental and Health Sciences, North-West University, Mafikeng Campus, Mmabatho, South Africa

**M. Mbewe**
Animal Health Programme, School of Agricultural Sciences, North-West University, Mafikeng Campus, Mmabatho, South Africa

**P. Veeramuthumari and W. Isabel**
PG & Research Department of Zoology and Biotechnology, Lady Doak College, Madurai, Tamil Nadu, India

**Uzay Gormus**
Istanbul Bilim University, Department of Medical Biochemistry, Turkey

**Ilhan Yaylim-Eraltan**
Istanbul University, Department of Molecular Medicine, Turkey

**Nur Selvi**
Ege University, Department of Medical Biology, Turkey

Printed in the USA
CPSIA information can be obtained
at www.ICGtesting.com
JSHW011407221024
72173JS00003B/448